the Waters

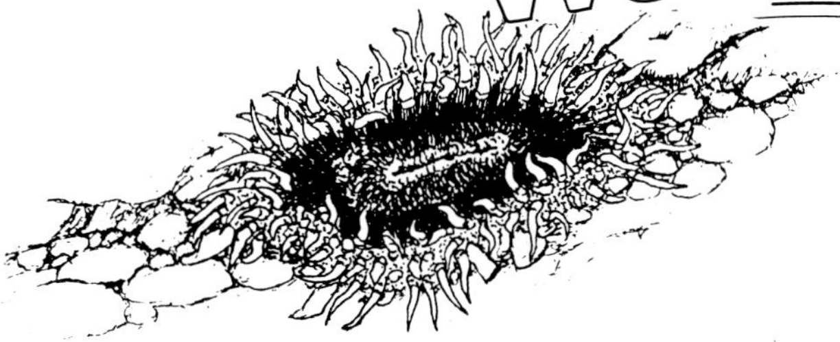

SUNSET OVER LAGUNA BAY

Atmospheric and cloud conditions often produce spectacular sunsets.

This photograph was taken in July 1978.

LITTLE COVE BEACH AND FIRST POINT

Sand is eroded by heavy south-east swells, leaving the protective granite boulders exposed. It slowly returns with calmer conditions, particularly in winter.

First Point was undeveloped in 1978.

NOOSA NATURE PART 1

the Waters

CECILY FEARNLEY

NOOSA NATURE
PART 1

The Waters

Printed by Sunland Printers Pty Ltd
Windsor Road Nambour QLD 4560

Set in 9pt on 10pt Palatino

Graphic Design by Julie Eijdenberg

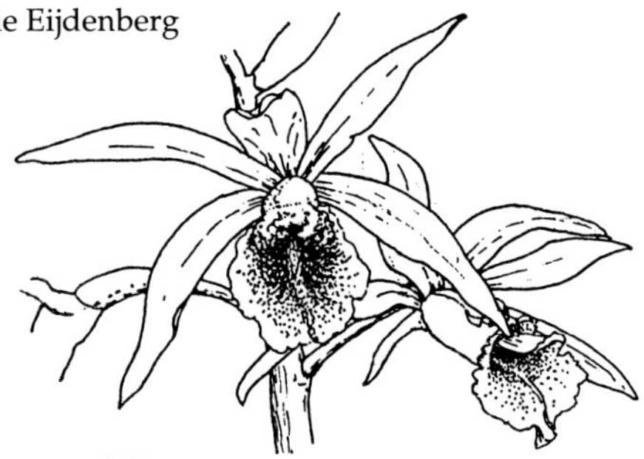

ISBN 0 646 00155 8

SET 0 646 00154 X

Preface

FOR NEARLY twenty years Cecily Fearnley has been writing weekly "Nature Notes" for the local paper, "The Noosa News". She illustrated them with her own pen-and-ink drawings. Now she has published the first of a series of books on the natural history of this beautiful region, which was still relatively unspoiled when she first came here.

The Noosa region is surprisingly diverse in the habitats it provides, and the forms of life it nurtures, both fauna and flora. From the richer soils and sheltered glades of the Noosa National Park, to its heath-like southern area, to open wallum and paperbark swamps, from the green tree-covered sand-hills of Noosa Heads to the river oaks bordering the Noosa's lower reaches and the subtle wildflowers of the coastal dunes, there is a huge Book of Nature open if only we have eyes to read.

In this volume Cecily has concentrated on life near water, from the seashore to the river and lakes. Each subject is illustrated with one of her detailed drawings, as well as a charming 'background' picture, and some of her own colour slides.

From small snails to large trees, she looks at her subjects with love and understanding, describes their habits and where you will find them. Through this and later books she hopes to share this understanding with a wider audience, so that with understanding will come the urge to protect and foster the wonderful natural world we inhabit here. Like Cecily Fearnley I have lived here for many years, and have seen with distress the bulldozing of whole blocks of living greenery, from fragile blossoming wedding-bush to irreplaceable old trees, and all the native life that went with them.

Let us hope that with the publication of "Noosa Nature" in this and later volumes, there will be more people in the area who care.

NANCY CATO

TITLE	SUBJECT	PAGE
Preface		7
The Story		12
A Bubbly Snail	Purple Ocean Snail	16
Dolphins Relate to Man	Dolphins	18
Floating Colony with a Sting	Portuguese Man-o'-war	20
Gannets, the High Divers	Australasian Gannets	22
Serpents of the Sea	Sea Snakes	24
Overnight Visitor	Green Turtle	26
Tideline Debris	Beach Shells including Abalone	28
Abstracts in the Sand	Soldier and Bubbler Crabs	30
Shore Patrol	Pied Oystercatchers	34
Not All Get Away	Saltwater Fish	38
Rocky Splash Zone	Littoral Zones	40
Building Dunes	Beach Spinifex and Other Plants	42
Palms on Stilts	Pandanus Palms	46
Flashes of Yellow	Small Yellow Butterflies	50
Life Between the Dunes	Dune Flowers and Insects	52

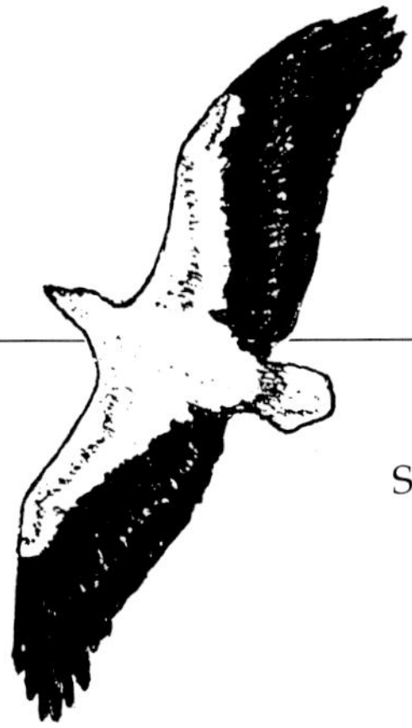

Contents

TITLE	SUBJECT	PAGE
Wandering Terns	Crested Terns	54
Estuary Life	Mangrove Dwellers	56
Sighing Trees	Casuarinas	58
Tree Termites and Uninvited Guests	Tree Termites, Birds, Goannas	60
Nature's Wall of Roots	Mangrove Trees	62
'Swampies'	Swamp Wallabies	68
The Long and Short of Local Tortoises	Local Tortoises	70
What a Jumper!	Rocket Frog	72
Serpent in the Reeds	Freshwater Snake	74
Regal Swinger	Royal Spoonbill	76
Water is their Element	Water Beetles and Bugs	82
These Plants Eat Flesh	Sundews	84
Nature's Helicopters	Dragonflies	86
Green Candles	Swamp Banksias	88
Lilytrotters and their Neighbours	Jacanas and Birds on Lake Macdonald	90
Scientific Names of Species		92
Index of Additional Drawings		95

STORM CLOUDS AT NOOSA

Wind, rain and hail in a fierce November afternoon storm is passing out to sea.

EARLY MORNING CALM AT TEWANTIN

Pelicans are a feature of both river and lakes, but do not breed locally.

NATURE AT NOOSA has evolved a delicate tracery of life. It is interdependent, and balanced perfectly to suit each particular habitat. This book aims to provide a glimpse into some of the many diverse life forms that live in the Noosa district, over, on, under, and beside both its salt and fresh waters.

Commencing at the ocean, I cite living creatures, resident and pelagic, that we can see from the coastline. Beaches and rocky shores are next explored, then mangroves with their specialised life forms that live only within that ecosystem. We move upstream, looking at vegetation and animals on the banks, reaching freshwater pools and their life forms, and finally large dams like Lake Macdonald.

I cannot attempt to tell all there is to know of a particular species, habitat or situation, but merely point out some facts that I find interesting, hoping that you do too. This is not a book on the botany or ornithology of the area. Many detailed books have been written by Australian experts in particular fields, books just waiting to be read.

Common names and terms have been used as much as possible, but at the rear of the book, you will find scientific names of items mentioned in the text.

Most of the information has come from my own notes, observations and experience. However, I have consulted as much relevant literature as possible. Many aspects of natural history have combined to produce this attractive, interesting region, and my aim is to encourage readers to become more aware of what is around them.

When I first saw Noosa in 1946, it already possessed a National Park (though not as large as today), it had wide ocean beaches, and its meandering mangrove channels were intact. Just two wheel tracks marked the road from the present Noosa Junction to Sunshine (then Golden) Beach. There was no coastal highway.

It was a pristine area, still in the form it had evolved over millions of years. Man had altered it little. Natural forces had shaped its coast, river and lakes. It was beautiful.

Initially, white man, finding it so, decided to stay and use the bountiful fish and huge inland forests. His presence caused great changes. Land was cleared for farming and housing, swamps were drained, mangroves filled, river channels altered, all of which affected the resident life forms in those places. My husband and I bought a holiday cottage in 1956, visiting every year.

The Story

When in 1972 we came to live at Noosa permanently I realised that many new residents were from far away, completely ignorant of the plant life they were destroying when they clear-felled their newly-acquired property. Some grass trees were hundreds of years old. Hollow limbs of scribbly gums gave shelter to squirrel gliders or provided lorikeets with nesting sites.

Some plants destroyed were very rare, and others (like the wedding bush and wax flower) could have produced masses of blooms every spring with no care or attention. These folk were replacing them with clipped lawns and European flowers with their constant need for maintenance, gardens that epitomised suburbia in any city in Australia. Both the 'look' and 'feel' of Noosa were being lost through lack of appreciation.

Thus I commenced writing weekly articles for the "Noosa News", covering all aspects of local natural history. I am still writing them eighteen years later, yet have only scratched the surface of the fascinating natural world around us.

I leave it to you to read on, hoping that the glimpse I give you of some aspect will cause you to open the door to a new interest, or seek to gain greater information in a specialist book on the subject.

Through knowledge and understanding, we will respect, appreciate, and retain habitats that could be lost through ignorance. Enjoy nature in the Noosa area, and pass on your enthusiasm for this beautiful place and wildlife to the children. The quality of their surroundings is in our hands.

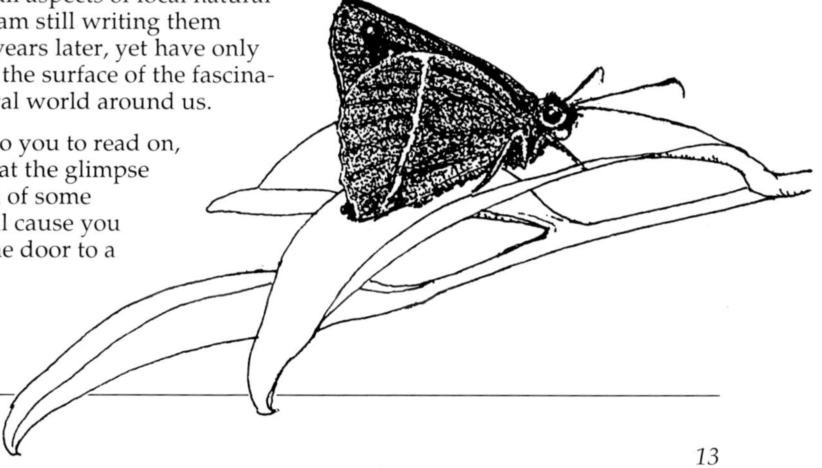

VELLELLAS, OR BY-THE-WIND SAILORS

Ocean creatures, they are moved by wind and currents, coming ashore anywhere.

These were photographed on the coral sand of Norfolk Island.

A STRANDED PORTUGUESE MAN-O'-WAR, OR BLUEBOTTLE

The blue floating air sac and interdependent groups of feeding, stinging and reproducing cells can be clearly seen. Its long stinging thread measured five metres along the North Shore sand.

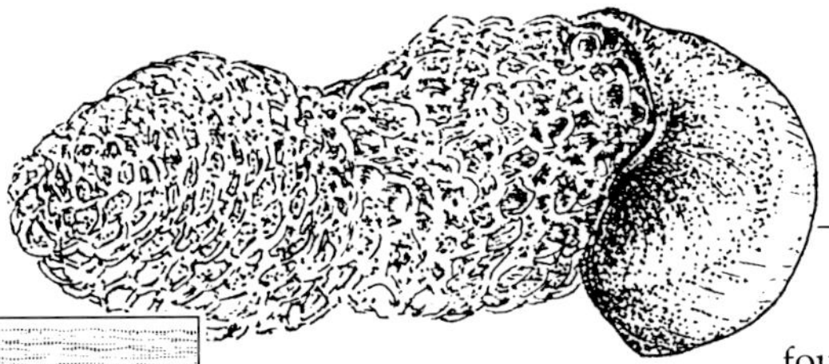

DO YOU KNOW that there is a snail that lives on the high seas? Yes, I said 'on', for the Janthina or Purple Ocean Snail lives and floats on the surface of the water, a true pelagic creature. Though both shell and body structure are basically similar to those of land snails, the Janthina cannot survive away from the ocean.

It is a true mollusc, a close relative to the molluscs that live among mangroves, on rocks, coral reefs or the ocean's floor, and move just above it like the scallops. All of them have two necessary requirements. The first is having a shell of some shape or size, and secondly having a soft unsegmented body.

Molluscs have been graded scientifically into five classes. First there are the chitons or coat-of-mail shells found in rocky pools at low tide. The second class covers single shelled snails, limpets, cowries, and also includes slugs. All bi-valves or two-shelled molluscs are included in the third class. These include oysters, which have one of their shells cemented to a rock, scallops, mussels, eugaries, cockle shells and similar.

Another class covers the elephant-trunk shell that lives burrowed into the sandy ocean floor and is seldom found cast up on the beach. Finally there is a class in which the mollusc has an internal shell. This includes the octopus, squid, cuttlefish and nautilus.

Most molluscs have evolved a muscular foot, a thickening of the body's undersurface. That is what helps limpets, for example, to cling so tightly to a rocky surface. You may think that an octopus does not have a muscular foot, but it is considered that its projecting arms are an adaptation of the foot, while the shell has shrunk and become enveloped by the soft body.

Janthina belongs to the single-shelled snails. How does it stay afloat? By means of a type of frothy bubble, formed from a secretion produced by the foot. This looks like a mass of foam, very soft and delicate. However the walls, though apparently thin and transparent are actually quite hard, similar in texture to a thin cuticle. As the mucus is produced, air is entangled and trapped before it hardens.

Thus this colourful, dainty, thin-shelled snail of the sea is able to float near the surface feeding on plankton, baby jellyfish, other small crustacea and even small ships' barnacles. When the Janthina in turn is sought for food it is credited with being able to exude a purplish liquid similar to the inky fluid of squids.

A Bubbly Snail

Protection also comes from its camouflage colouring for it floats upside down with its pale blue 'top' showing from below. Then when a fish looks up and sees a Janthina it is almost the same colour as the sky. Conversely its upturned dark blue 'base' when viewed from above by a hungry bird looks remarkably like the ocean below.

Being completely at the mercy of winds and currents, the purple snail is often blown towards land. Sometimes hundreds end up stranded along our ocean beaches. What a wonderful opportunity we have then to admire and examine this beautiful mollusc!

Other examples of floating ocean life will probably be stranded too. Small dainty harmless Velella or 'By-the-wind Sailors' may be there in their thousands.

Nearby could be Physalia or Portuguese Man-o'-wars with their tough translucent blue air bags for flotation and clusters of different tentacles. Velellas are like Physalias in having similar groups of tentacles, though in miniature form. With them on the beach could be a line of tiny blue *Glaucus* Nudibranchs. They are a type of sea slug with fringed extensions like limbs.

All these ocean dwellers feed on microscopic plankton and move at the mercy of wind and currents. So after days of strong onshore winds, take a walk along our ocean beaches. See what the tide has left stranded. The flotsam may even include a Janthina with its bubble float still attached!

THERE ARE MANY simple pleasures to be enjoyed at Noosa and one I know is absolutely free. That is to sit beside Laguna Bay at sunset and enjoy not just the colours of the sky and its reflections on the moving waters, not just the rumbling background wave noises and surge of the breaking surf, but both of these with an added bonus.

Who does not thrill to the sight of a school of dolphins as their fins break the surface? They arch, then slide gracefully out of sight once more. The quiet ease of their evening appearance blends perfectly with the mood of the day, something appreciated and enjoyed by resident and visitor alike in the half light.

Dolphins have a special relationship with man, and anyone who has visited an aquarium where these animals are performing or on display will understand what intelligent creatures they are. People of the Pacific Islands have legends and stories of friendships between man and porpoise, and we here in south-east Queensland have a similar tale to tell.

Anyone who has read Tom Petrie's "Reminiscences of Early Queensland" will remember his account of the porpoises of Stradbroke Island. The aboriginal tribe living on the island would watch for schools of mullet to appear offshore in winter, then would go down to the bay and smack the water's surface with the broad side of their boomerangs.

The sound must have carried quite some distance for soon dolphins would appear and commence herding the school shorewards. There was soon food in plenty for both man and dolphin, the result of a close relationship between two intelligent creatures.

Its body structure has slowly changed over millions of years. Once a land animal, it now dwells constantly in a marine environment but retains vestiges of arms and legs in its skeleton.

One aspect of interest is its means of communication. We hear it as a series of loud clicks. These can vary a little in length and frequency, and it is thought the timing and intensity of the 'clicks' may act as a personal signature.

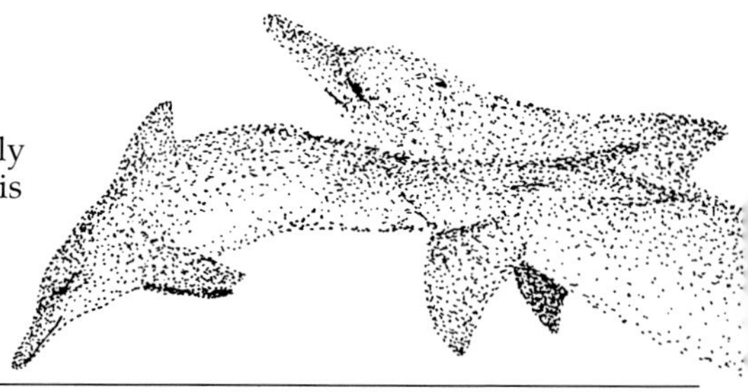

Dolphins Relate to Man

Some scientists believe that their chief function is to echo-locate food, a technique also used by flying bats to catch insects and avoid night collisions.

I read in Heather Angel's book, "Monsters of the Deep":

"Dolphins appear to produce their clicks by extruding a fine stream of bubbles out of their blow-hole, possibly with the aid of the valves and associated little sacs. The shrieks they can produce seem to come from their larynx in much the same way that human speech is formed. The front of the dolphin's cranium has flat forward facing surfaces that act as a reflector of the sound, and the pad of fat overlying the skull on the front of the face acts as a sound 'lens', concentrating and directing the acoustic signal forward.

For instance if a dolphin is swimming straight towards a hydrophone it will pick up the sounds loud and clear, but the intensity diminishes considerably as soon as the dolphin turns slightly to one side."

Scientist are still trying to analyse the abilities of dolphins, but for the average person the obvious is most interesting. Both the sudden leaps they make in Laguna Bay for no apparent reason, and their zest in surfing the waves around Dolphin Point catch our imagination. After the initial shock and surprise of the boardriders, what a thrill it must be to have a dolphin join you on a wave!

Everyone is sad when one is caught in the shark net offshore. It happens even now. I remember some years ago a young dolphin was struggling to escape the net. An adult dolphin was splashing and leaping all around, trying desperately to attract a human's attention. It did, and two locals set out in a rowing boat, found the problem and at last managed to release the young one.

As they worked the adult kept nudging the boat, as if urging them to hurry. Alas, they had not come soon enough, and the youngster died near the end, drowned.

The adult nudged the lifeless body, circled the boat a few times, then disappeared. Was it saying thanks for trying, or angry that man's nets had taken its young? Who knows? If baited hooks were set to catch sharks instead of nets then the graceful friendly dolphins would be able to swim safely in our bay and continue to give us those simple pleasures.

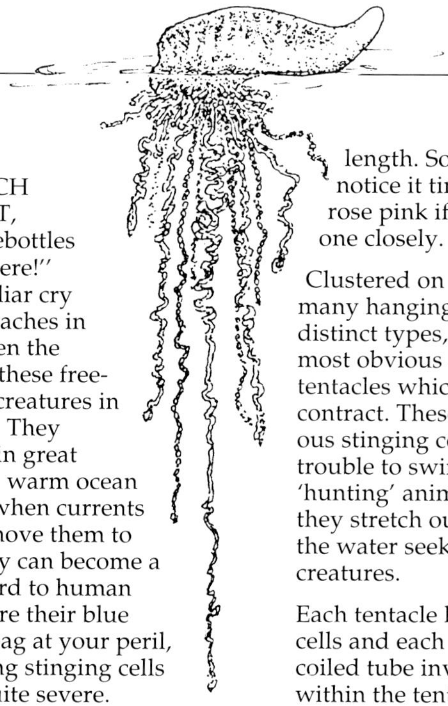

"WATCH OUT, bluebottles are everywhere!" What a familiar cry along our beaches in summer when the winds blow these free-floating sea creatures in to our shore. They usually live in great shoals out in warm ocean waters, but when currents and winds move them to the coast they can become a painful hazard to human surfers. Ignore their blue floating air bag at your peril, for the trailing stinging cells can prove quite severe.

The Portuguese Man-o'-war or Bluebottle belongs to the jellyfish family, but it has no swimming bell. Most jellyfish have a circular skirt which moves them along through the water by muscular contractions.

The Bluebottle is propelled by a gas-filled bladder catching the wind, causing it to move over the surface. Most bladders on our beaches are between 5 and 10 centimetres long, but some overseas specimens have been reported as being 20 cm!

Though of a translucent blue colour, the bladder has a narrow crest of slightly darker blue running its full length. Sometimes you will notice it tinged with purple or a rose pink if you care to examine one closely.

Clustered on its undersurface are many hanging appendages of three distinct types, all a darker blue. The most obvious are long thin trailing tentacles which expand and contract. These contain the numerous stinging cells which give all the trouble to swimmers. They are the 'hunting' animals of the colony, for they stretch out and move around in the water seeking fish and small sea creatures.

Each tentacle has numerous stinging cells and each of these has a fine coiled tube inverted into a pocket within the tentacle. This contains a toxin so that when contact is made, the tube is thrust out of the pocket rather like a finger of a rubber glove being turned out. The poison then travels down the tube and penetrates the fish (or human) skin. Tiny barbs help the poison-carrying tube to stay in position and also makes it harder to rub off.

While stinging cells are doing their work the tentacles hold fast the prey, and by contraction bring it closer to the main colony. Then the feeding zooids take over, each with a sucker-like mouth on the end of a little tube. This is applied to the fish, and food is drawn up the tubes and

Floating Colony with a Sting

sent off to nourish all parts of the colony.

The remaining group of zooids merely have the task of reproduction. Thanks to the efforts of the other three, they do this very well. Witness the great numbers we occasionally find in the water and then stranded on our beaches.

Don't think that being stranded on a beach is anything unusual: it isn't. That has been happening on Australian beaches for the last 650 million years! David Attenborough's book, "Life on Earth", tells of fossils found in the ancient Edracara Sandstone of South Australia's Flinders Ranges. Though these ancient rocks were thought at first not to hold fossils, they have found some that have the ripple marks of a sandy beach on the bedding plane. They noted there were a few flower-like impressions, some small, others to 8 cm across. Paleontologists determined they were the marks of jellyfish that had been stranded on an ancient beach, baked in the sun, then covered by the next tide.

They were beautifully preserved. Sixteen species have been discovered, and some of them have gas bladders like the present Portuguese Man- o'-war. It certainly makes one wonder how many millions of years they have been in our oceans. Though the stings are so painful, there are two ocean dwellers that seem unaffected by them. One is a gaudy fish named Nomeus that lives amongst the tentacles, impervious to the toxin. It feeds on the Bluebottle's victims. (Anemone fish do likewise amongst anemones.)

The other is the Loggerhead Turtle. Apparently this reptile does not mind being stung, for fishermen have seen the turtles, their eyes swollen and nearly shut imperturbably munching their way though patches of Portuguese Man-o'-wars. I am sure there are many swimmers along our beaches in summer who will wish that there were many more Loggerhead Turtles around!

NOOSA IS A beautiful place in winter as well as summer. June to September brings warm sunny days with clear blue skies and calm seas which merely ripple and glide onto the wide beach. This is the time for us to welcome many southern visitors. Not only human visitors come to Noosa, avian ones arrive too.

Look across Laguna Bay and you will see some of them wheeling gracefully through the air, long white wings with distinctive black ends and rear edges proclaiming they are Australasian Gannets and they have arrived for the winter season. Occasionally non-breeding birds may call in or fly past in any month, but winter is gannet time at Noosa.

Keep watching the birds. You will see that, after cruising leisurely between 10 and 20 metres above the waves or swooping low, one will suddenly twist and dive vertically. Its keen eyesight can detect fish from a range of distances. The streamlined head and body are kept in a straight line, with the wings slightly closed to accelerate the drop as well as maintain course. When close to the surface the wings are trailed back to form an arrow. You can see the splash from shore.

Studies of the birds have revealed they do not spear their fish like the darters, but catch them from underneath. Usually fish are swallowed under water, but some birds bob up with one still in their bill. Quick manipulation ensures the fish goes down head first, and the gannet is ready for another search, catch and eat.

The dive can continue as long as ten seconds before reappearance, for the bird has evolved a great help in maintaining buoyancy. Boobies and gannets have both developed a system of inflatable air sacs under their skin. Just before they hit the water these sacs inflate to cushion the impact shock and assist in the bird's return to the surface. After resting and floating a few minutes our gannet is off and flying again.

Another special device helps the gannet in its spectacular dives. Its pointed bill is so streamlined it has no external nostrils! How does the bird breathe? Secondary external nostrils have developed at the gape, the angle of the mouth. These have a flexible flap over each nostril, closed by water pressure when the bird dives. Neat, isn't it!

A wingspan of 1.7 metres and a body length of about a metre make it a striking bird. The wings are white, with black primary feathers (the wing tips) and black secondaries (along the trailing edge). Central tail feathers are blackish-brown, which helps you distinguish

Gannets, the High Divers

this bird from the Red-footed Booby whose whole tail is white. Australasian Gannets have colour, a lovely golden yellow on their head and neck, plus a light grey-blue bill. However these can be seen only at close quarters.

Immature birds occasionally seen in summer are a brownish-black, with spots of white. That gives them a speckled appearance, giving way later to the adult's dapper plumage at about two years. Apparently they breed only after six or seven years.

Like so many other water birds, they like the companionship of their own kind when breeding. Over the centuries they have chosen definite sites where they congregate in hundreds, but none is near Noosa. Two known rookeries are off the Victorian coast, and four off Tasmania. Other sites are in New Zealand, and the birds cross the Tasman Sea. Young birds banded in Hauraki Gulf, New Zealand were found at Northhampton, Western Australia nine months later. Obviously young birds are great travellers.

Gannets have been studied by ornithologists for some time. They noted that if the seas are too rough and winds greater than Beaufort Scale Force 4 they can not see fish clearly. Both waves and water are too confused, and the birds go hungry. No wonder they come to Noosa in winter.

As you relax and watch these graceful birds diving, remember they may have come all the way from New Zealand to enjoy our weather and fishing.

ARE THERE sea snakes swimming off our beaches? Yes indeed, and occasionally after a cyclone or bad weather one is stranded on the beach, exhausted or dead. If you do find such a victim, you have a wonderful chance to see how cleverly nature has adapted this snake to its life under the ocean waves.

Sea snakes generally are just a group of hollow-fanged venomous land snakes which have evolved special abilities to live almost all their life underwater. Fish are their main diet, and because they need a quick kill their venom is highly potent. Short teeth penetrate scales and skin only a short distance, so after biting it does not let go but keeps its jaws 'chewing' to make sure the venom is well injected.

Sea snakes are not aggressive towards man, but are apparently extremely curious. I am told they are liable to swim quickly towards a diver and coil themselves around his legs or arms. In trying to disentangle himself the diver is liable to be bitten.

The first thing you will notice is their specially flattened tail, a vertical flattening rather like an eel, and they progress though the water in an eel-like manner. While land snakes need wide belly scales for land-gripping propulsion, these are no longer needed by their aquatic relatives. Those scales have degenerated into scales not much larger than the others.

As all sea reptiles need oxygen, sea snakes have to rise periodically to the surface for a breath of fresh air. Notice how their nostrils are situated on the top of their head, not on the sides as is the case with land snakes. They also are specially modified with a flap of skin which acts as a valve, closing after the intake to keep air passages waterproof.

You may have noticed how land snakes have a centre-front opening to their mouth through which their valuable sense organ, the tongue, is able to protrude with the mouth still closed. This is formed by the rostral shield being almost hollowed out. In sea snakes the shield has a prolongation downward which fits into a notch in the lower jaw, thus sealing the mouth against the entry of any sea water. Clever, isn't it!

Some sea snakes do spend some time in among mangroves and nearby land. Their ventral scales are wider than the purely aquatic relatives. The Banded Sea-snake is the commonest of these. It has a body ringed with many alternate

Serpents of the Sea

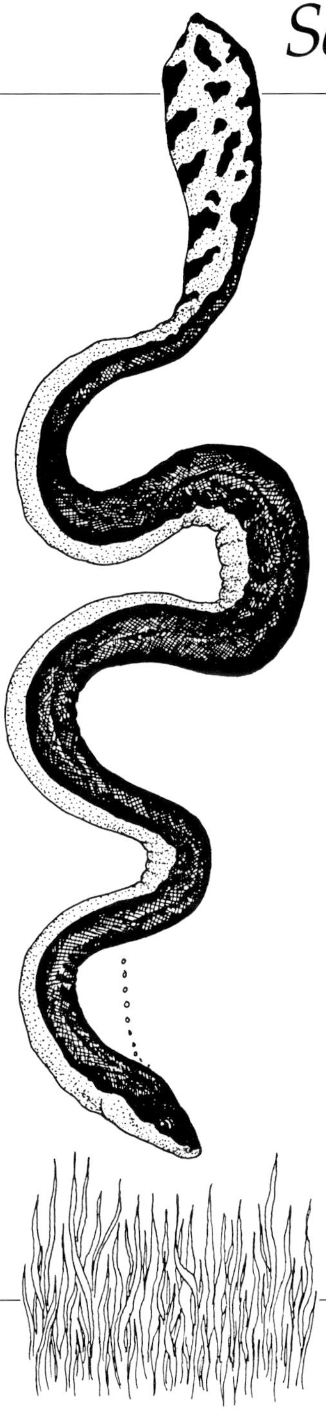

bands of cream (or white) and blue-black, with a cream front edge to its head. With twenty-one sea snake species recorded in Australian waters (mainly in the tropics) only five are regularly found off our Noosa coastline. Generally they range in colour from dull grey to brown, with or without darker bands.

Probably the commonest and certainly the brightest in colouring is the Yellow-bellied, which grows to about a metre long. It has the top half of its body black, the bottom half a bright yellow to soft brown with the tail mainly yellow with black blotches. This snake has a world-wide distribution, living all its life at sea. It produces live young, able to fend for themselves immediately. That is usual for sea snakes, but the Banded lays up to a dozen eggs on land, close to mangrove areas.

Life histories and habits of our sea snakes are the least known of any reptile group. Living in the ocean restricts greatly their availability for study. Perhaps the enquiring minds of present marine students will later unravel many mysteries of sea snakes and their lifestyles.

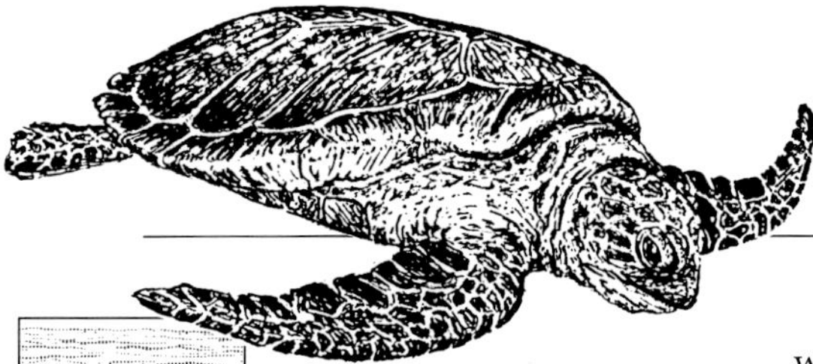

EACH DAY'S high tide leaves new pages of the sea's never-ending saga, ready for people to read and appreciate. Though some days will have little change on the page, others will contain exciting records. They could range from the arrival of an unusual shell to the finding of strange footprints in the sand. Imagine the excitement of one visitor, who witnessed baby turtles hatching!

Any summer night between October and March turtles come ashore for egg laying. Mon Repos beach near Bundaberg is a favourite site for both Loggerhead and Green Turtles. Some of the latter also come to Sunshine Beach for nesting. Almost every year a few of the adults' tell-tale flipper tracks are found leading to and from the low dunes.

Years of research by Q.N.P.W.S. officers have disclosed much new information about turtles through tagging, weighing, measuring, plotting nest sites and recording sand temperature. However, more has yet to be done.

They have found that only female turtles visit the beaches; the males remain swimming just beyond the surf. They arrive offshore about August, remaining until November with mating taking place in the ocean near the nesting site. Female Green Turtles reach quite a size before maturity, the smallest carapace measuring 93 cm for an 80 kg turtle, the largest a long 1126 cm for a huge 200 kg reptile. Imagine lumping that weight up the beach with flippers!

In captivity a turtle can grow to the smallest nesting size in about ten years, but a free turtle grows more slowly. Research on growth rates suggests that it will be over fifty years old before it begins to breed. Not only that, but a female appears to breed for only one season. How important then these nesting sites are!

The female may have swum many hundreds of kilometres to reach the right beach, but to date it is not known if each turtle returns to the beach where it was born. Suffice to say instinct tells it when it is safe to come ashore, usually after dark, and preferably from one hour before to two hours after the high tide.

Turtles are timid creatures for all their size, and are easily 'put off' by lights or strange movements and return to the water. But when all is quiet they haul themselves from the shallows, up the beach to the foredunes, beach spinifex, shrubs and even trees. Some turtles are

Overnight Visitor

known (through tagging) to return to lay at fortnightly intervals, and one was recorded eight times! If she had laid the average clutch of 112 eggs each time, surely some of her progeny will live to maturity!

Once the female reaches a flattish point on the dunes above normal high tides, she begins to scoop away the loose dry surface sand. A 'body pit' forms as large as herself, and having reached the damp firm sand beneath she then carefully scoops out a vertical hole with alternate hind flippers. The right depth (about 40 cm) and right temperature (25°C to 35°C) means the white spherical eggs will have the right warmth and protection.

Soft-shelled like all reptile eggs, they are liberally coated with mucus. Then sand is scooped back on top of them, tamped down, and more added until the hole is completely covered. Finally she digs her way forward, throwing sand around to disguise the site. After about eight weeks the first egg hatches, though they may take longer at Sunshine Beach which has cooler sand than Mon Repos.

The first hatchling's movement stimulates the others to do likewise and together they dig to the surface. This is a case of unity being strength. After two to seven days they reach just below the surface where they stay awaiting nightfall. Only about 5% emerge during the day, when they are at the mercy of gulls and hot sand, for the sun can dehydrate and kill them.

Once the sea is reached there are all the hungry sea creatures to evade. Every time they surface for air they correct course to swim directly away from land. The ocean's horizon light guides them, and any closer brighter land lights will disorient them. On Heron Island they often flip their way to the lights instead of the ocean.

When Sunshine Beach hatchlings flip their way down the beach and into the surf let us hope many will survive to return and lay their eggs in fifty years time.

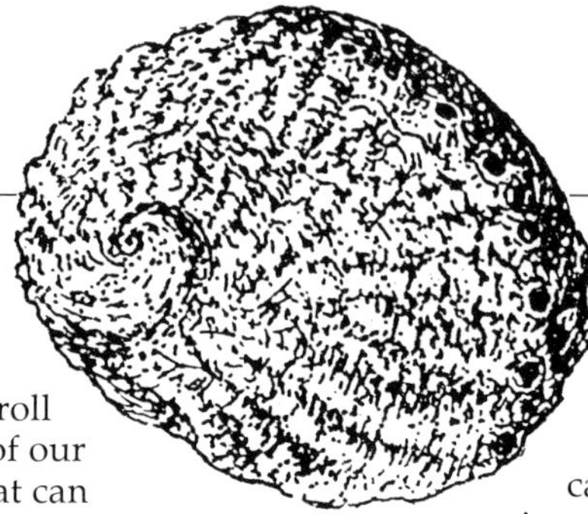

TAKE A leisurely stroll along one of our ocean beaches. What can be more relaxing to visitor and resident alike than to amble along an open expanse of beach and feel a clean fresh breeze in one's face, blowing away any cobwebs and tensions in one's mind. All this is with the background noise of the surf's dull roar.

Waves are continually breaking, running up the beach, pausing, then starting to soak into the sand before they can be dragged back to the next one. Sometimes the wave brings something with it, too heavy to be floated back, and it is left high and dry on the sand, the flotsam and jetsam of the sea. Depending on the tide and wind directions an amazing variety of items, both natural and man-made, are left behind.

Wander along Sunshine Beach or the North Shore and investigate the high water mark. With so much vehicular traffic now, North Shore shells do not stand a chance of remaining intact for long. A few years ago before the 4WD cars and trucks were so popular, a day's beachcombing became absorbing, and one was sure to discover beautiful shells.

The variety of sea discards is surprising if one really looks carefully and does not ignore everything but the biggest and brightest. If summer winds are from the north-east there may be days when the beach is covered in Bluebottles. Other days tiny delicate blue Nudibranch become stranded, or numerous Violet Sea-snail shells are left on the sand. Next to white Cuttlefish backbones lie different pieces of broken coral, and lumps of pumicestone from some volcanic source.

These days man helps swell the sea's discards by adding his own. Plasticware, large and small, joins pieces of timber and odd boxes tossed from passing boats. Even less welcome are light bulbs and bottles. But it is the marine life remains that have a right to be there, and undoubtedly shells are the most interesting. The study of shells, or conchology to give it its correct title, can be a life-long interest. If you become curious and wonder just what might be the name of a strange and beautiful shell you have found, then I advise you to buy a good reference book.

A flattish, nobbly shell I found not long ago was quite different from the usual types. On looking up my reference book, Coleman's "What Shell Is That?", I found in the Abalone section that it is called the Warty Ear Shell.

Open and only slightly domed, it has no internal spiral though the outside suggests the remnants of one. An intriguing series of raised holes is peculiar to this type of shellfish, and on turning it over, a glistening silvery pink and green inner lining is revealed.

The pinkish-brown outside surface is ridged tightly by low sharp lines in a circular direction, while large raised ridges radiate roughly from the apex cone. That gives rise to the 'warty' description, for the ridges are quite irregular.

Again "What Shell Is That?" gives information that this shell

"lives on reefs from low tide level down to at least 50 metres. It inhabits crevices, ledges and caves. Specimens have been recorded from deep water off Southern Queensland, but apparently it is not common here, being at the end of its northern range. One of the largest abalone, its food consists of drift algae scraped off the surrounding reef surfaces. Besides man, its predators include stingrays, crayfish and sharks."

I find it very interesting to discover what sea creatures live off our beaches, if only from the shells cast up on the shore. I hope you do too.

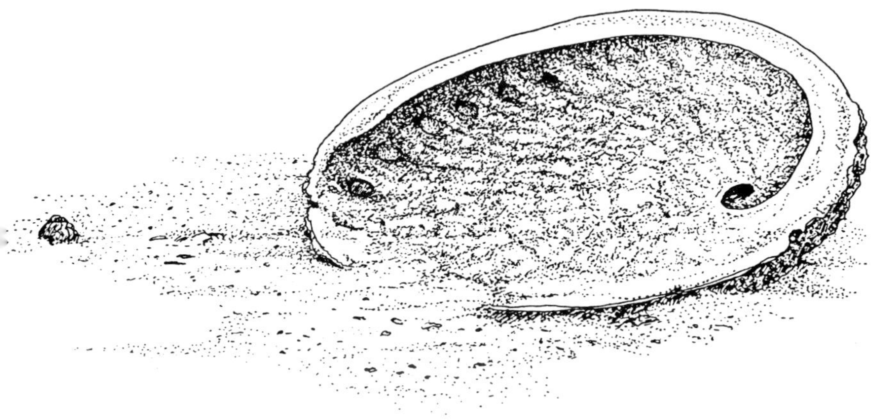

ANYONE walking our beaches and sand banks at low tide is sure to have seen tiny balls of sand surrounding a small hole. "Ah, a crab's hole" is the immediate thought, and you are quite right. You would be lucky to see the crab actually at work; it is usually just the handiwork you notice. The crab has fled back down its hole!

If you walk on an estuary sand bank, then you may see sand balls left by soldier crabs. They are gregarious creatures, unusual crabs in several ways. For example, they walk both forwards and sideways, whilst most crabs walk sideways only.

Since Noosa Sound was created from Hay's Island and the Spit area formed, we do not have the large estuary sandbanks that were home to millions of these comical blue and white crabs. The armies have gone, but platoons, companies, or perhaps a battalion can still be found on a quiet bank.

They feed on minute organisms left by the tide, and if disturbed – what a rustling sound as they move away. Sometimes they walk over each other in their haste! To escape they dig into the wet sand with a distinctive sideways corkscrew motion, legs on one side scooping the sand while those on the other circle backwards. This is quite unusual in the crab world.

If the sand is very soft and moist they can disappear in as little as five seconds. Sometimes they remain just under the surface ready to suddenly appear when danger has gone, but generally speaking most burrows reach depths of from 15 to 25 cm.

Out on the ocean beach it is a different crab that makes those round sand balls: the Bubbler Crab. Low, sandy-coloured and semi-transparent, it is much smaller than the Ghost Crabs which also inhabit the beaches. Like them it is capable of great speed sideways when scurrying back to its burrow. Most crabs seem to hold themselves high as they survey the beach scene, but the Bubbler has a much lower carriage.

Though they're found on ocean sand beaches and able to live in exposed areas, they are more numerous in quieter, more protected bays and gentler slopes. Like the estuary's Soldier Crabs, they too rest at the bottom of their burrows until the tide has ebbed and the sand surface has become exposed.

Emerging quietly, the tiny crab commences feeding. It picks up

Abstracts on the Sand

sand with its chelae, and several pairs of mouth appendages busily remove all-important organic food matter that has been left by the tide. It then rolls the used sand into pellets, leaving them on the beach surface.

As it progresses further and further from its burrow it makes shallow trenches radiating from the opening of its vertical hole. This is one creature that does not camouflage its lodgings, for the round sand pellets and shallow radiating trenches are a dead give-away. Actually these trench patterns are of interest to photographers who see beauty in the round pellets, especially in the low afternoon light.

Too often we view things from our full height, but if you come down to almost the crab's level and use a magnifying glass it is surprising how attractive these pathways are. However there is no need to get on hands and knees to appreciate this tiny crab's handiwork. Just take a stroll along the beach at low tide.

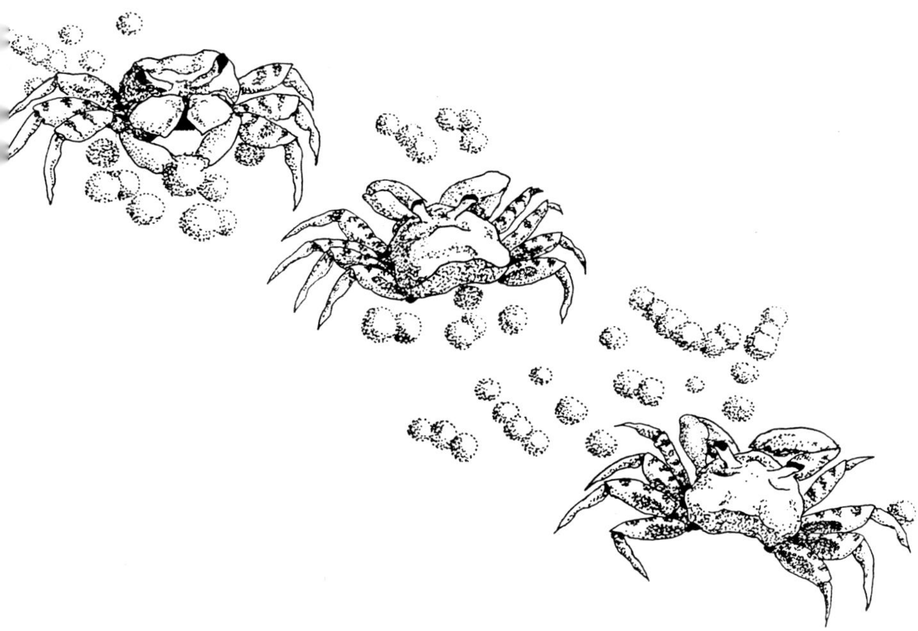

SAND BUBBLER CRAB HOLES

The characteristic pattern left by these small crabs at low tide reflects their different sizes. Estuary sand is darker than ocean beach sand, due to its added decomposed mangrove leaf materials.

BLUE SOLDIER CRABS

At low tide, estuary sandbanks erupt with small armies of blue soldier crabs. They feed on stranded microscopic organisms. When moving together, their numerous tiny feet make a rustling sound.

DRIVE ALONG THE North Shore beach and somewhere before you reach Double Island Point you will see a pair of Pied Oystercatchers. Alas, there are only two or three pairs left now, but years ago when beach vehicles were few and far between it was a different story. Then there were many pairs of these short, dumpy black and white birds living quiet, undisturbed lives along the beach.

Each pair appeared to patrol a particular length of shoreline, though there was no obvious boundary at either end. The birds seemed to agree by mutual consent where territories met and kept to their own during the breeding season of August to January.

During high tide you may still notice them standing on low dunes below the coloured sands, preening their feathers. Then, as the tide drops, their search for food begins. Standing with bright red legs in shallow water, their bills probe down into the soft wet sand feeling for a eugarie shell. Finding one, the bill is quickly jabbed between the open valves into soft white flesh before the eugarie can snap shut.

A seagull's bill is not strong enough to handle the muscle power of a eugarie (or pipi as it is also called). The Oystercatcher's bill is about 7 cm long, higher than it is broad, and shaped like a chisel. It can be jabbed in quickly and thus paralyses the shell's occupant before the two closing side muscles can work. Not only eugaries are eaten: these birds also search for worms, sea snails and other marine life of the intertidal zone.

In non-breeding months boundaries are relaxed slightly, and you may see a family of four or five birds standing on the beach together. The young stay around their parents for about nine months, then go off and form small groups in estuaries such as Tin Can Bay. In some places around Tasmania very large flocks are known to congregate. It is not until the birds are about four years old that they pair off and in turn commence breeding.

When resting out on a sandbank, Oystercatchers usually stand on one leg, heads turned round and bill tucked neatly between their wings. If you walk towards them they will

Shore Patrol

look round, watch you with their bright red eyes, then walk away when they think you are too close. Continue walking, and they will go on, and on, then eventually fly. After a quick prancing run along the sand their wings unfold for low flight while uttering their high-pitched alarm call 'hee-eep'.

Flight reveals an interesting pattern, for the black wings have a startling white bar, and the previously hidden rump is revealed as pure white. Usually they turn soon after takeoff and fly back, landing where you first saw them.

One would think the birds would have learnt how to contend with vehicles by now, but apparently not, for not long ago I saw a lone Oystercatcher standing near the body of another which had obviously been knocked by a car. Perhaps it had foolishly tucked its bill into its back for forty winks.

A Pied Oystercatcher's nest is merely a scrape in the sand well above high tide mark. The two to three eggs are olive-grey with dark blotching giving good camouflage. If danger comes near to eggs or chicks, the parents can turn on a good imitation of a broken wing. They also try to fool you by pretending to sit in another place. If all else fails they will even attack the intruder, using a snapping bill, dives, and striking wing tips.

Around rocky headlands and on stony beaches you may see another oystercatcher, the Sooty. Although the same size and shape as its cousin, this bird is all black except for bright red eyes, bill and legs.

It searches for mussels and limpets on the rocks, and also of course for oysters. It too jabs quickly into a bivalve before it can snap shut, and prises off limpets and other molluscs, smashing them against stones.

A few have been seen around Noosa's headland over the years, but the Oystercatcher most loved here is the beach dweller, the Pied. If it learns road sense, and if there are still eugaries on the beach, it will be with us always.

DUNE CREATION (right)

A strong south-east wind is carrying dry beach sand. Dune plants are catching some, while a drift forms across the beach access track.

GOATSFOOT CREEPER (far right)

Large cloven leaves and delicate one-day blooms characterise this sand-binding plant.

BEACH SPINIFEX FLOWERS

These later develop spiky seed heads which are blown tumbling across the sands to scatter and germinate, helping to bind further this fragile area.

SUNSHINE BEACH DUNE PROTECTION

Beach Protection Authority fences prevent people from trampling and lying on plants needed to hold sand and re-build eroded dunes.

FISHING has been a challenge to man ever since the first hooks and lines were invented. Nowadays fishing still presents a challenge, whether it be with the basic equipment of hook, sinker, and line wrapped around a glass bottle, or the very latest in custom-made lures and hooks with almost transparent yet strong nylon line gliding from a fibreglass rod and spinning reel.

Noosa's amateur fishermen have a choice of three fishing situations: rocky headland, surfing beach or placid river. Each site attracts fish according to their feeding preferences.

Rock fishermen are happy to clamber down the cliffs of Noosa National Park so they can tempt Jew, Tailor, Trevally, Bream and Morwong to snap at their bait. Many fish come in close to feed on Cunjevoi that live attached to the rocks at low tide level and below. Small fish, crabs and molluscs seek food and shelter amongst the seaweeds there and are in turn sought by larger fish.

Each type of fish is tempted best by a particular bait and every experienced fisherman has his own 'surefire' theory as to what to use, when, and how to use it.

One of the most beautiful of fish is the Golden Trevally. A young specimen has golden sides flushed with silver, its head and snout tinged with green, and ten to twelve narrow vertical stripes along its sides. It is a determined fighter, but after being landed the colour quickly fades. Not just death, but increased size through age also causes fading.

The maximum weight for a Jewfish is 65 kg, but happy is the fisherman who returns with a 10 to 15 kg catch. At night these fish swim along the surfing beaches in gutters close to shore. Surf fishermen cast into the gutters with very long rods. A finger is kept on the line, feeling every movement caused by waves and current, patiently waiting for the different tug that means not just a nibbling crab but a fish taking the bait. Then the battle is on.

In the half light of early morning and sunset, Tailor are tempted by metal spinners which are reeled in close to the surface. They must look remarkably like a lonely pilchard to the hungry fish which snaps it up. Beach gutters hold the same species which feed around the rocks. This time they seek other molluscs, worms, and more small fish in the

Not All Get Away

shallower beach waters.

Sand (or Summer) Whiting are found in estuaries, sandflats and along ocean beaches. There too Trumpeter Whiting are caught in winter. Bream and Tarwhine are found and caught around our waters most of the year. Both are great fish to eat, and crowd into our river estuary to breed. They need shallow sandbars at the river mouth to lay their eggs. After hatching, the fry seek the quiet shelter of mangrove backwaters where they can grow before venturing out into the open sea. That is one reason why mangrove areas are essential near the estuary; they provide a continuing supply of Bream to eat later!

Sand Flathead are caught near the beaches, while Mud Flathead prefer the river, especially from May to July. In daylight they are rather sluggish, spending time partly buried in the sandy bottom, waiting for a small fish, crustacea or other food morsels to pass close by. Then they snap it up.

Calm winter seas are often ruffled by schools of fish moving in the bay. Watch a flock of feeding terns, how each flutters, wheels and dives into the milling pilchards. Splashes come from Tailor chopping into any fringe fish, so the pilchards are constantly trying to keep within the shoal!

Sea Mullet also pass by in large schools during winter, but they are neither as large or frequent as in years gone by. Commercial fishermen post a lookout on the dunes so the beached boats can set out and net a good catch. The same also applies to Tailor.

Many other interesting edible fish are caught around our coastline and in the river, each giving its captor a quickening heartbeat, pumping adrenalin, and sometimes sore muscles. By keeping to legal limits in size and numbers, and ensuring the mangroves and sandbars remain, we should be able to catch a good feed for generations to come.

ANYONE who wanders around the rocks along the Noosa National Park shoreline at low tide will be sure to see many interesting sea creatures at different levels.

Just as vegetation changes as one climbs up a high mountain, so different depths along the shoreline are home to different sea creatures. These bands are called littoral zones, and in some places are clearly seen.

The supra-littoral zone is highest. There, where the spray of high tide keeps them wet, one finds tiny smooth blue-grey periwinkles. Seeing them clustered together in groups seeking dampness in slight depressions, one wonders how they can survive so long without drying out. Those rocks can become quite hot in our summer sun.

Another small univalve shell is found there too, a periwinkle with circles of ridges and the impressive name of *Nodalitorina*. Often there is no common name apart from the general word 'periwinkle'. Although seeming to live high on the rocks these molluscs must descend to lower levels to breed, for their young are part of the many life forms of plankton. However, as they cling to the hot rocks, one does admire their ability to survive in a difficult environment.

The next belt is indeed the littoral zone, with several sub-zones according to the tides and associated plants and animals. Surfaces exposed to ocean swells with surges of spray differ in depth of zone to the quieter sheltered rocks with small differences in wave depths.

Acorn-barnacles are found in the top zone. These strange sea creatures start life looking similar to a young prawn, then change as they grow, their shell plates later becoming firmly attached to a rock. Surrounded by other hard sharp-edged plates, the centre two open when covered by the sea so feathery feet can protrude. These wave around in the water urging microscopic food into its mouth. When the water recedes the two plates close, protecting the animal for almost half of each twenty-four hours.

Large barnacles are usually found on the upper limits and small ones lower, at about mid-tide level. These tiny ones can be closely

Rocky Splash Zones

packed together and give the rocks a soft grey colour. One recorder has counted 12,500 on a square foot of rock surface. Other larger barnacles may be mixed between these types depending on the average wave size and the rock's aspect.

Below the mid-tide level, one usually finds a hard encrustation of white tube-worms called *Galeolaria*. You may think they look a bit like interwoven coral, but no, they are tube-worms. They go down to about half a metre above low tide mark, and there one starts finding the Cunjevoi.

Every fisherman knows the Cunjevoi or Sea Squirt. Watch the rocks at dead low tide and you will be fascinated by the squirts of water that suddenly shoot up from these creatures. Each one is a brownish, rough, leathery column about 12 cm high, with two openings on top. One inhales the water and food particles, the other ejects the water after feeding. Its flesh is dark red and a popular bait for fish.

Among the Cunjevoi are often found large brown seaweeds which extend to well below the surface, into the sub-littoral areas. It is here that many different seaweeds can be found, and a great variety of sea creatures hiding in them.

Found among the rocks in the barnacle zone are many small living molluscs with shells. Observe but do not take them, and remember to always replace any stone you may overturn, or you may kill all sheltering creatures. Crevices and shallow pools harbour strange oval-shaped Chitons, and short-tentacled Anemones which lie camouflaged with pieces of broken shell attached to their muscular lips.

There is so much to understand of the creatures that live among the rocks. Each has its own degree of resistance to wave action, heat and dehydration, so each selects that part of the littoral zone that suits it best.

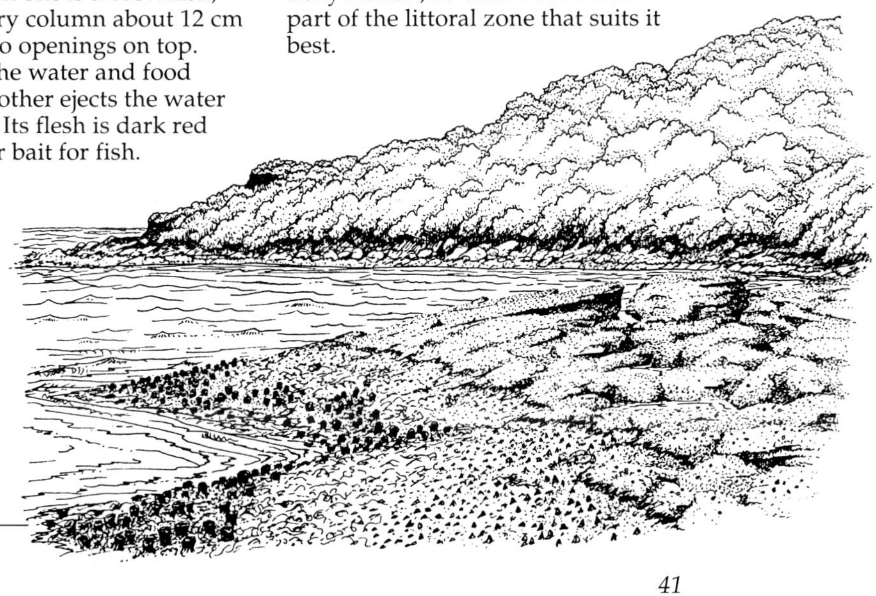

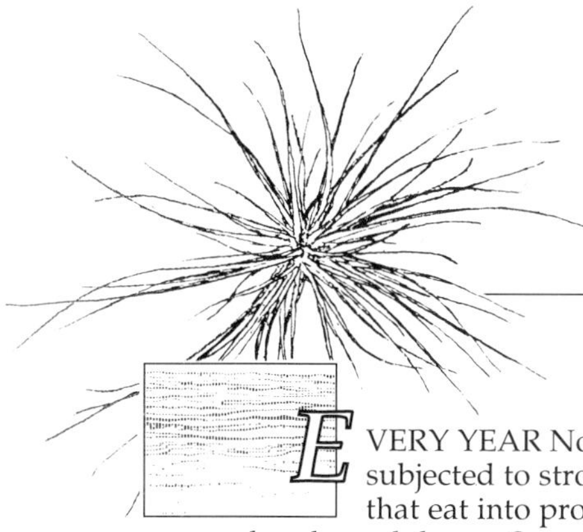

EVERY YEAR Noosa is subjected to strong waves that eat into protective beach sand dunes. Summer cyclones are always possible, but seldom are they extreme here.

Generally speaking, they pass well out to sea, merely leaving us a legacy of wind, rain and then good surfing waves, particularly around Noosa National Park. However in some years (such as during February 1972), a series of passing cyclones can result in days of huge waves that cut into the beach, foredunes and even the face of the high dunes as well. When the sand goes, soft 'coffee' rocks are revealed, and at several places, particularly opposite the mouth of 'First Creek', one can see at low tide the stumps of large melaleuca (paperbark) trees still standing in the surf.

Pondering over the reasons why they are there, it would appear that at some time in the not so distant past these trees had been growing in a sheltered swamp. One can only wonder how long ago the salt water broke through their protective dune to surge in, drown them, then snap off their trunks.

Nature has a wonderful ability to adapt and heal its self-inflicted wounds. When sand is taken away by wave surges, it goes out only a short distance, and gradually returns when the weather is calmer.

Each wave then brings a little more back and normal winds pick up the drying sand to blow it further up the slope. Who hasn't had their legs stung by this wind-borne sand?

Gradually the beach is returned, the low foredunes re-formed and built up against the scarred high dunes. Vegetation holds them together, and though it sometimes takes years, the open beach returns to its former contours.

The vegetation's vanguard is the Beach Spinifex. This is the grass-like plant that sends out those long pale green trailers across the dry beach. Their thin leaves wave softly in the breeze, becoming the first living thing to halt the wind-blown sand.

In early summer clusters of tiny flowers develop into spiky seedheads which break off and are blown along the beach. Thus seeds are scattered, later producing new plants to bind the sand still more. Although they look unimportant, those tumbling spiky balls are vital for Nature's beach regeneration programme.

Close behind the spinifex, and doing their best to hold the sand that blows past it, are Pigface, Goatsfoot, Beach Bean, Scented Fan-flower and other salt-resistant plants.

Building Dunes

Pigface is not an endearing name and one wonders why it was ever given to this useful plant with fleshy, long three-sided leaves. The large pink flowers (yellow in a different species) are up to 8 cm wide with numerous narrow petals. Sometimes the flowers' colours are so intense that they almost glow in the sun. The plant spreads low over large areas of the dunes, thickly in protected valleys, as well as surviving in most exposed positions on both dunes and rocks.

Goatsfoot runners are usually nearby, the long prostate trailers having large cloven leaves which help trap more errant sand, and roots that help bind it together. The 4 cm wide mauve flowers look rather like those of the Morning Glory Vine.

Trees play their part too. In the vanguard are the Beach She-oak or Casuarina, Coastal Banksia and Pandanus. It is surprising that the She-oak, such a fragile, delicate looking tree, could withstand strong winds and the salt that comes with them. Yet all three trees grow well in these trying conditions, protecting the next line of timber that is less salt-tolerant. This zone includes Paperbark and Bloodwood trees.

Nature's healing forces for beach replenishment are constantly at work. The saga of dune restoration and the consolidation process is a fascinating story. We can read it every time we visit the open beach, but please give Nature a chance. Avoid walking on the plants; keep to the pathway. Those growing tips are building a beach for you to enjoy.

TERNS AND GULLS ON THE NORTH SHORE

Crested Terns are the most numerous of Noosa's five tern species. Here a few Silver Gulls join them when resting at low tide. These birds roost in large numbers on estuary sandbanks at night.

FRESHWATER BEACH POOL

Some freshwater creeks draining onto the ocean beach are stained a rich tea colour. The stain is caused by decaying vegetation of peat swamps and heathlands.

WALK INLAND from the beach and one of the first trees you will find is the Pandanus. It loves to grow within sight and sound of the sea. Whether it grows on a sand dune or coastal cliff you will find this unusual tree spreading its branches with a crown of leaves at each tip.

Screw-pine is another name for it, and notice how those leaves are arranged in a spiral fashion. The 'screw' part may be relevant, but it is not a true pine.

Yet another common name is Walking Tree, for the base of the trunk sprouts numbers of thick, stiff, slow-growing roots for a metre or more up the trunk. These are likened to legs or props anchoring it to the ground as extra support against wild ocean winds.

What you do not see are finer roots extending long distances through sand or rock fissures. When cyclonic winds cause waves to smash in and erode beach dunes, pandanus trees are sometimes left perched above a great sand cliff. Then these long trailing roots are exposed, showing just how far the roots have had to travel in their constant search for nutrients.

Fruit grows at the end of the branch, looking rather like a round pine-apple nestling amongst the leaves. It is spherical, about 25 cm across, forming a large cluster of about fifty pieces. Though green at first, it ripens to orange, then crimson. Each kernel has a stringy covering, and when it falls to the ground needs suitable soil contact and weather to germinate. If it grows around the headland, the tree is often surrounded by a thick mat of grass, or rocks and boulders, so few kernels manage to reach soil and fulfill their purpose.

This is not the Breadfruit tree that Captain Bligh of "Bounty" fame was sent to bring from Tahiti. That tree is quite different, and belongs to the mulberry family. Its fruit, although of the same size and shape as the Pandanus, has a rough rind surrounding a mealy pulp, which both looks and feels like new bread.

William Dampier discovered the Breadfruit tree in Polynesia way back in 1688, and the only other place it grows freely is the Carribean Islands, where it was originally taken to feed slave labour.

Our Noosa Pandanus loves salt air to blow through its leaves. It needs lots of water in summer and little in winter, so our climate suits it perfectly.

All Pandanus belong to a large complex genus found world wide in

Palms on Stilts

tropical and sub-tropical coastlines. However, some species grow in rainforests, swamps, on river banks and on plains, as happens up in Arnhem Land. No matter where they grow, each species has the same distinctive type of leaf. It is spirally arranged, is long, tapering, and armed with small spines along its edges and under its mid-rib.

Though easy to propagate from seed, these thorny leaves make choosing a site in your garden particularly important. An older tree (with its dead leaves removed) gives dense shade which is so welcome on a hot day. That shade is greatly appreciated when it is beside a surfing beach.

Thousands of people have walked along Noosa National Park's coastal walking track. They have enjoyed the melaleucas, the banksias, the casuarinas and especially the unusual Pandanus. When next you walk that path, sit for a while beneath a Pandanus tree and really study it. Remember its special features. Then look about you. You will be seeing a part of our natural world unchanged by time. Look beyond at the ocean and headlands and drink in the colour.

Through preserving such scenery and vegetation in a National Park, we are ensuring that thousands in the future will enjoy it all as we do now.

NOOSA NATIONAL PARK
WALKING TRACK

This 1957 photograph shows both a pandanus tree and scene across the bay that present vegetation has almost completely obscured.

DEVIL'S KITCHEN HEADLAND,
NOOSA NATIONAL PARK

Taken in 1973, this photograph shows the rugged headland still relatively bare from a fire some years before. Vegetation is now general on the headland. Lion Rock is to the right.

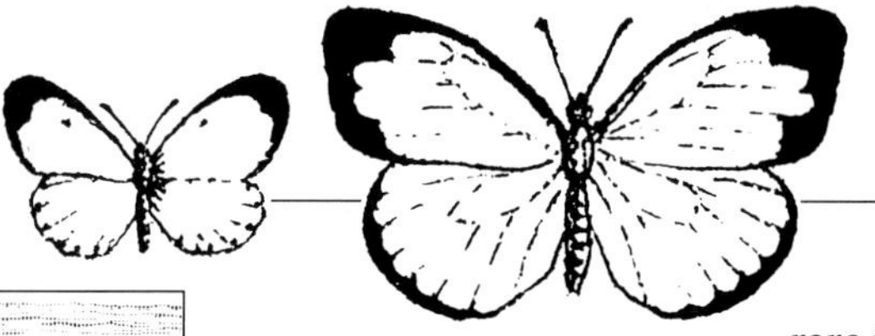

SO OFTEN in this world we tend to look for and appreciate only the most spectacular, the biggest, fastest, longest and so on, with regards to both human values and the world of nature generally. Thus we tend to overlook the attractions of smaller things which often have a grace and beauty of a completely different scale.

Butterflies are a specific instance. The attention they receive is so often in direct proportion to their size. A small butterfly alighting on the grass or at a tiny flower is accepted as part of the scenery. A large one makes us notice and exclaim with pleasure, yet the smaller may possess delicate patterns of intricate design and be extremely rare. Because of size and low flight it passes unnoticed.

That is usually the fate of many tiny 'blue' butterflies, some only as wide as a thumbnail. However, everyone at some time has noticed small bright yellow butterflies flying fairly close to the ground.

As you sit on a sand dune admiring a pigface flower's vibrant colour, what joy when a small brilliant yellow butterfly visits it. Which butterfly is that? It could be one of four similar species at Noosa, one common, two seldom seen, and one rare in this area. All have wings of bright yellow with black edgings, varying only in size and positioning.

Largest and most frequently seen is the Common Grass Yellow. Like all butterflies of this genus, it has sulphur-yellow wing colouring, is small, and flies close to the ground with a characteristic erratic movement. Our Common Grass Yellow with a wingspan of 45 mm is particularly attracted to flowers, and although it has an undulating flight it settles frequently.

If it settles on the Pigface flower notice its wing patterning. The forewings have a broad black terminal band, with the inner edge deeply indented about one third of the way along. Only a narrow black line edges the rear wing. Both wings are yellow underneath, with the female adding variable brown spots or blotches. Always look carefully at the forewing's black areas for they tell you which of the four species you are looking at.

Only a fraction smaller at 40 mm is Macleay's Grass Yellow, having the same brilliant yellow on its wings with black markings. It is a rare butterfly at Noosa. Its black edging is on the point of the forewings only, the hindwings merely having terminal spots. This butterfly often has a pink under-surface with one or two curved brown streaks, but

Flashes of Yellow

they vary considerably. The same general black markings apply to the smallest of the four at 25 mm, the Small Grass Yellow, sometimes found here. Its apex is more rounded, and it has none of the pink of Macleay's.

Last but not least is the butterfly with the strange name of No-brand Grass Yellow. About 37 mm across, it has the usual black edging but adds a narrower strip around the rear wing edge than the largest of the four, and a stronger yellow beneath. It is sometimes seen flitting over the grass.

Male sex brands are found on the first three species, but not on No-brand, hence its name. Sex brands are grey or pinkish patches of specialised scales found under the forewing near its base, and in others also opposite on the hindwing's upper surface. You will never see them if the butterfly is flying!

All four butterflies, so similar in low undulating flight, are often dismissed with only a cursory glance as being identical. Their golden beauty as they flitter low over beach dunes or through open forest is like captured sunlight on a grey day. Small as they are, they contribute so much to making any bush walk or even a lazy afternoon in the garden a place of quiet pleasure and delight.

When next you notice a yellow sprite, pause and see if you can distinguish which one it is. Is it common, uncommon, or even rare? If it eludes you and flutters on, do not be disheartened. Perhaps it will settle longer next time.

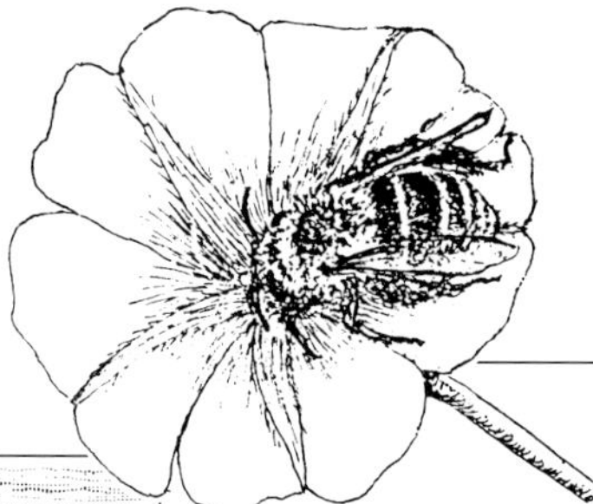

DUNES facing the ocean beach are constantly being shaped by the wind. Dry sand blows up from the beach, is trapped by plants, and builds the foredune gradually higher. The front is shaped naturally to deflect wind up and over not just the first but also the second or main dune itself.

Thus between both dunes and beyond a depression forms, and in those sheltered spots thicker vegetation grows. Horsetail oaks rear their fine waving branches, coastal banksias with short cream 'candles' become established, and pandanus palms grow well. The undergrowth becomes dense, insects move through and feed on it, and birds on them. A habitat is established.

Find such a gully and sit for a while observing nature.

Goatsfoot creepers are there with their delicate pink convolvulus flowers. They open each morning but by mid afternoon the wafer- thin outer petals have curled into the tubular centre. Bees must be on the job early to gather pollen from these blooms! Goatsfoot creepers can be easily identified by their large 'double' leaf which has a vague resemblance to the hoof of a goat.

Commercial European Bees are active around the dunes. They usually belong to an 'escaped' hive, which could be in a tree hollow up to three kilometres away. One often sees bees drinking at a freshwater soak on the beach.

Small black stingless Native Bees also visit flowers in those sheltered spots. Sprays of the pea-shaped Beach Bean are soft pink, while light blue Scented Fan-flowers are small, five-lobed, and nestle down amongst the plant's rather fleshy leaves. Nearby could be white Beach Stackhousia, and tiny pink Emelies.

Pigface, with its circle of narrow pink petals around a yellow centre, may attract yet another bee, one of the Carpenter or Solitary bees. These bees do not belong to a hive, but after mating the female constructs brood chambers either in underground tunnels, or hollow wood, or in stems, depending on her species. There she places pollen and honey sufficient to support each of her eggs to maturity. There are thousands of different native bees in Australia, mostly solitary, and much work has yet to be done identifying them and studying lifestyles.

Many other trees, shrubs and low plants flourish in well established dune gullies. Growing around and over any fallen branches could be the Twining Guinea Flower with its large yellow blooms. All stages of the flower's growth have beauty of shape and colour. The silverish blue-green nut-shaped bud opens to a wide flat golden bloom, then forms bright red seeds contained in five large bronze sepals. Strange to say it does not have an attractive scent, except perhaps to flies!

Small brown ants run up and down the stems and trunks searching for nutrients, while small skinks scurry under leaves as they in turn search for tiny bugs, beetles, moths and spiders to eat. Assassin Bugs lie in wait for flies or anything that may inadvertently come close so they can make a quick grab. It is a case of eat and not be eaten.

A Wolf Spider may move across the sand between leaves. At first glance you may think she has a woolly abdomen, but look closely and you will find she is carrying her family on her back. Dozens of tiny spiderlings cling to her, sometimes climbing down when she stops, for they feed on her capture too. When she moves they quickly hold on again.

Nearby could be a Sand Wasp finding a suitable spot in the sand for a nesting hole. After digging it (rather like a dog), she searches for food. It could be flies or other insects, or spiders, depending on the speices of Sand Wasp. Paralysing but not killing her prey, she carries or drags it to the hole, sometimes nipping off its legs if they make it too heavy. Then, laying an egg on the paralysed flesh, she covers the hole with sand and flies away. In spring, all the young wasps emerge from the sand, having finished their food supply, spun their cocoon, and pupated.

They in turn may become a bird's meal, for a Rainbow-bird, Drongo and Cuckoo-shrike all enjoy a wasp dinner. Honeyeaters eat insects too, though mostly seek pollen and nectar. When the banksias and melaleucas are in bloom they arrive, the Blue-faced, Brown, Scarlet and White-cheeked competing with Rainbow and Scaly-breasted Lorikeets amongst the trees. Searching for insects in the gully below may be both Wrens and Scrub-wrens, and (if sufficiently dense) you may even hear the crack of a Whipbird!

These sheltered gullies between fore-dunes and behind main dunes are a world to themselves where the life and death struggles of insects can be observed every day.

"LOOK at all the gulls!" is an exclamation often heard around the Noosa River estuary. Visitors may be speaking of the flocks of white birds neatly resting on sandbanks near the river mouth, flying out of the river in the early morning or returning in the late afternoon, or even diving for their tucker into a school of pilchards out in the bay. A white trail of terns and gulls following a fishing boat back over the bar is a beautiful sight that often calls for exclamation.

To call them all gulls is wrong, for most of the birds are Crested Terns. Though Silver Gulls are most numerous in many other areas, at Noosa Crested Terns far outnumber the gulls.

A general rule to help distinguish between gulls and terns is that gulls have a more rounded and deeper wing silhouette compared to a tern's angular shoulder and thinner, more pointed wings. A gull's bill is shorter and thicker compared to the tern's pointed bill, while its legs are longer than a tern's in proportion to the body.

The Crested Tern has a greyish-white body, darker above and on the wings. The crown and nape are black, but during non-breeding months white streaks and speckles appear on the crown. Its white forehead may also be speckled then, but in spring and summer it is a clear white with a clean-cut edge against the black crown.

Do not be confused by the name if you can see no crest. The long nape feathers are normally smooth and flat, but when a bird is disturbed or agitated, especially at nesting time, they are erected to form a distinctive crest. The best identification of all is the colour of the bill, easily seen in flight or when resting. A pale yellow bill with a greenish tinge shows it is an adult, while an immature has a greenish-yellow bill.

If amongst a resting flock of terns you notice a large bird with a bright red bill, it will be a Caspian Tern; if it has a black bill you can be fairly sure it is a Gull-billed Tern, though sometimes Common Terns are seen in our river; if it is small, dainty, and has a cream bill with black tip it is the Little Tern. Other visitors come to share the river sometimes, so there is always the chance a Roseate, Lesser Crested, or a White-winged Black Tern may call.

However, the Crested Terns are always on the river, out in the bay, and up into the lakes. They have a long sweeping wing action and generally graceful flight. They bathe in shallow water, then dry off and preen out on the river banks.

Wandering Terns

When fishing, either as a flock or as individuals, they dive straight in. Observers have noted that a small fish is gripped just behind the head. The tern ascends holding it in its bill, then swallows it in mid-air. When a bird is feeding young, it flies purposefully, sometimes as much as three kilometres with the fish in its bill.

Crested Terns breed in close colonies, their shallow scooped-out nests only a metre apart, and usually on a rocky headland or sandy beach. The closest known breeding colony is on Lady Elliot Island. The birds range for hundreds of kilometres up and down the coast, so the terns here today may be well away next month. One bird banded in south-east Tasmania was found at Noosa seven months later, a distance of 1,950 km direct! That was the longest recorded movement for this species up to 1986.

Careful observations, as well as banding, increases our knowledge of Australian birds and their movements. So if you are on a boat cruising close to a Crested Tern resting on a channel marker post in the lakes, look well at the bird. Perhaps it will be wearing a leg band placed there years before, a very long way away.

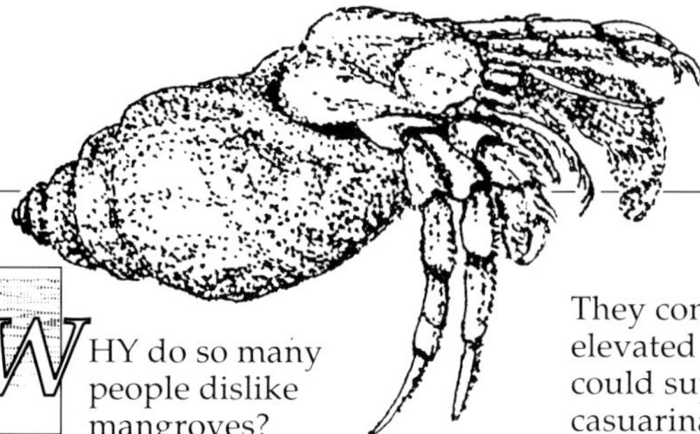

WHY do so many people dislike mangroves? They are not just muddy smelly places. They are areas with a myriad of specialised plants and animals which have adapted to live in a difficult environment where land and salt-laden water meet.

Those of us who knew Noosa Heads when Hay's Island existed remember the quiet winding mangrove-lined channels with their good fishing. While drifting in a boat between walls of dense mangroves with their arching, inter-twining oyster-covered aerial roots, one was in a different world. Those arching roots formed a protective barrier which absorbed destructive wave action, while behind them the sandy mud rose into islands growing other mangrove species.

Noosa's seven species, particularly the tall Grey Mangroves, are the primary stabilising force that turns shifting sand banks into a consolidated environment. Mangrove roots are shallow but spread wide. Some species have vertical extensions up through the mud called pneumatophores. Lenticles grow on them like small knobs, special adaptations which absorb oxygen essential to healthy plant growth and not obtainable in the thick mud.

They consolidate so well that elevated areas like Hay's Island could support land trees such as casuarinas, acacias, and even some eucalypts and grasses. However, much of the island was covered with water during king tides.

Keyser Island still has similar vegetation to the original Hay's Island.

True mangrove mud flats seem distasteful to many humans who are unable to walk on them easily, or utilise them for building, farming and so on. They think of them as wet muddy places, breeding sandflies and mosquitoes, and thus consider them quite useless.

Yet they are so rich in animal life as to be the most prolific life support system of any soil. Unfortunately most of the life forms are too small for us to see, and so are ignored. Both plants and animals have specially evolved to live in this quiet water environment, where drifting currents distribute a rich store of rotting debris from the tree clad land.

Zostera weed beds grow in many mangrove channels and are highly nutritious to several forms of marine life. These sea grasses provide a haven for fish fry in their early stages of growth. The saying, "Destroy breeding grounds and you destroy the fish!" is closely bound up with mangroves.

Estuary Life

Another interesting factor is their ability to cope with varying salinity levels. Most ocean creatures find it impossible to survive in brackish water, yet mangrove trees and their associated life not only tolerate long periods of 'fresh' but indeed thrive.

With the area so rich in organic matter there is an abundance of animal life, divided roughly into the types that seek it below the surface, like worms, and those that feed on the surface. Hordes of Hercules Club shells, whelks and many crabs come out and deliberately pick up particles of film left on the mud's surface.

Hercules Club molluscs are common in all mangrove mud flats, and although not considered edible by today's standards, were thoroughly enjoyed by the aborigines of this area. Midden heaps reveal much regarding their tastes in food.

Small Hermit Crabs move everywhere over the mud's surface with their borrowed homes, while larger and more agile unencumbered crabs move in and out of their holes. The mud walls are reinforced with sand and are surprisingly strong. The whole surface seems perforated with the holes, whatever the state of the tidal flow.

Some birdlife has also become specialised for mangrove living. Mangrove Kingfishers with their green-blue back, black crown and pure white front are readily noticed. The Mangrove Gerygone warbles its sweet musical song, unexpectedly attractive for such a plain brown-grey bird. Mangrove Herons and Mangrove Bitterns can often be seen along the river banks seeking crabs and other food amongst the roots and mud. Other waders like Spoonbills and Ibis enjoy feeding close to mangroves but are not 'specialist' mangrove birds whose life depends on the continuing existence of these areas.

We all like the natural aspects of Noosa, and must be careful that enough of its mangroves remain to mature fish fry, and all other completely dependent life forms.

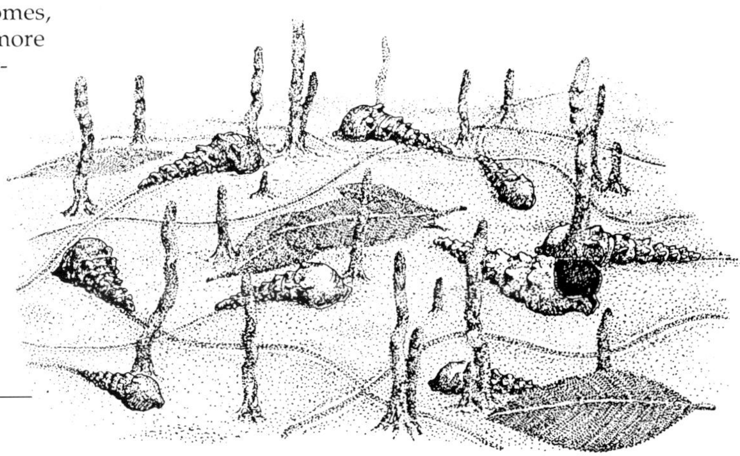

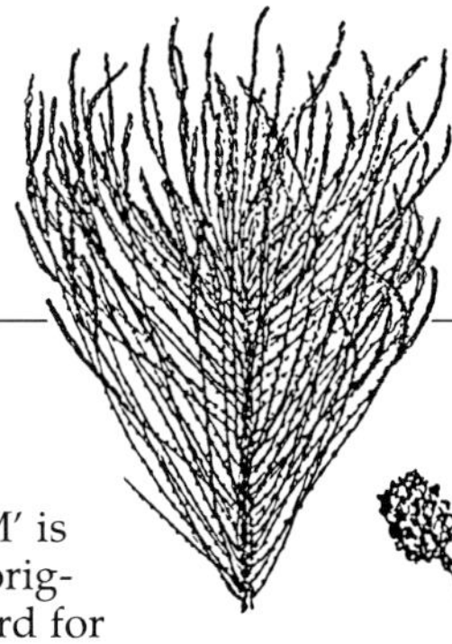

'WALLUM' is the Aboriginal word for the large banksia tree with its great yellow candles of flowers, and it has become accepted now as describing collectively all the low coastal country where these trees flourish.

Other plants found in these areas have also been given that common name, so we have the Wallum Callistemon, Wallum Boronia, and the Wallum Zieria. Yet another is the Wallum Oak, a she-oak or casuarina which in winter displays a spectacular mass of flowers around Noosa. Do not look for flower heads like the masses of Bloodwood blossoms we see each January, or the cream blossoms adorning paperbarks or melaleuca trees also in winter. If you are able to take a trip up the Noosa River in June, you will find the air heavy with the scent of paperbark honey, bees working furiously, and honeyeaters satiated.

No, casuarina flowers are quite different. Male flowers are minute, but in their millions cover all the tips of the oak, giving the whole tree a soft reddish-chocolate hue. A gentle breeze moves the branches and you hear a soft shooshing sigh.

Some people say that is why the casuarina is called a 'she'-oak. Others maintain it is because the timber grain looks like oak but is much more brittle. Perhaps you have heard another reason given.

All she-oaks, be they on sandy wallum areas or growing on inland river banks, have several distinctive features. Their barks are rough, dark-coloured, and have an abundance of slender, faintly striped, jointed green twigs that are often mistaken for leaves. When the pieces of needle-like branchlets are pulled apart, a row of minute teeth is seen at each junction. These are all that remain of leaves on these trees, and are used for identifying the different species.

Male and female flowers are quite distinct, and found on different trees. The male flowers are easily seen as they consist of a single stamen each, grouped in very slender orange or brown spikes at the end of the twigs. Female flowers grow closer in, on older branches. They appear as dense clusters of reddish or yellowish (depending on the species) stigmas protruding from small knobs along these branches.

When these have finished flowering a small woody cone develops, consisting of densely compacted seed capsules which eventually split open to set free extremely small seeds. The cones are very sharp and prickly, especially to bare feet if you walk under the trees!

With the Wallum Oak the branches usually curve upwards near the ends, and the green twigs are erect and spreading, quite different from the drooping Coastal Casuarina which grows along the beach front and has its twigs drooping downwards. One tends to think of them as leaves, but remember they are only soft, pliable twigs.

There are four species of Casuarina commonly found in the Noosa-Cooloola region. The Swamp Oak *(Casuarina glauca)* is a conical or rounded tree growing to 20 metres; the Forest Oak *(C. tortulosa)* has very slender drooping branchlets and grows from 15 to 20 metres in moist gullies or on hillsides. The other weeping foliage belongs to the Coastal or Horsetail She-oak *(C. equisetifolia var. incana)*. It grows as a small spreading tree on sand dunes and cliff faces, enjoying salt spray on its branches.

Finally the Wallum or Black She-oak *(C. littoralis)* can form a tree up to 12 metres high in a forest situation, but out on the open windswept coastal heaths grows as a shrub less than a metre tall.

As you move around the Noosa coastline, up its river and into its hills, keep your eyes watching for these brown tinted trees in winter and spring. You will be surprised at their number. Nearby will be the female trees with their coyly hidden flowers, the 'she' she-oaks.

Tree Termite

HOW PLEASANT it is to glide along the upper Noosa River. So many different trees, vines, shrubs, ferns, flowers and reeds to notice, with birds and insects giving added colour and movement.

Notice also the large number of tree termite nests that cling like great muddy lumps to the sides of tree trunks. Some stand out even more when the dead tree has been partly broken by a storm and stripped of limbs. Always there is a long mud-covered connecting trail from the arboreal mound to the ground, for the termite's body is very susceptible to injury by sunlight. It moves in darkness as much as possible, and any break in the covered trail is quickly sealed by mud.

These elevated termite mounds do a double job. Not only do they provide a snug home for their builders, but they are also homes for kingfishers and tree goannas. In the Noosa district four kingfishers compete for these specialised building sites. They are the Kookaburra, Sacred, Forest, and Mangrove Kingfishers.

A particularly large, strong termite nest is needed for the bulky Kookaburra, but the small Forest Kingfisher has a wide choice. This brilliant blue and white bird prefers them for its nesting site rather than a horizontal tree stub, spout, or broken-off limb which it sometimes uses. In fact a Forest Kingfisher might migrate locally to areas where tree-termite mounds are available for its use.

When its four to six rounded white eggs hatch between October and January, the chicks keep up persistent trilling most of the day. This quickly brings an observant person to notice the termite nest with its hole.

Sacred Kingfishers usually nest by burrowing into a bank a metre or so before making a nesting chamber. They also use hollow limbs in eucalypts or mangroves, but sometimes decide on a termite nest high on a tree. They too like a room with a view.

One researcher noted that a particular tree-termite nest had been used by Sacred Kingfishers for thirty

nd Uninvited Guests

years, while another nest had been used for twenty years.

After a new family has flown, the termites often seal the nesting chamber's entrance. If so, then next spring when the prospective parents arrive they take turns in drilling the opening. The hardened mud is pierced from repeated raps of their closed bills. Not quite as difficult as for woodpeckers in the U.S.A. making holes in wood, but all termite nests are hard. It probably gives the birds quite a headache.

Lace Monitor lizards, or Tree Goannas as they are also called, are a hazard for young kingfishers. They easily climb trees to search for and eat young birds. Often they too use the termite tree-mounds as a home for their eggs. Their sharp claws tunnel into the mound before the lizard deposits between eight and twenty rather elongated, parchment-like white eggs.

After a goanna's visit the termites rush around sealing over the entrance, and do not seem to mind the eggs at all! The constant warmth from decaying organic matter inside the termite colony incubates them perfectly.

The young lizards take about eight to ten weeks to develop, then literally cut their way out of the egg with a sharp 'egg tooth' on the tip of the snout. This falls off soon after birth, but not before they have also scratched their way out of the sealed-up wall which was probably softened by fluids from the broken eggs. These brightly banded miniatures in yellow and black scramble down to the forest floor where life for them begins in earnest.

So, when you see a dull grey-brown termite nest up on a tree, look carefully for a tell-tale hole, or a section freshly sealed. The hole may suddenly wear brilliant colour as a kingfisher flies in or out, or – just maybe – you will have the thrill of seeing a baby monitor emerge!

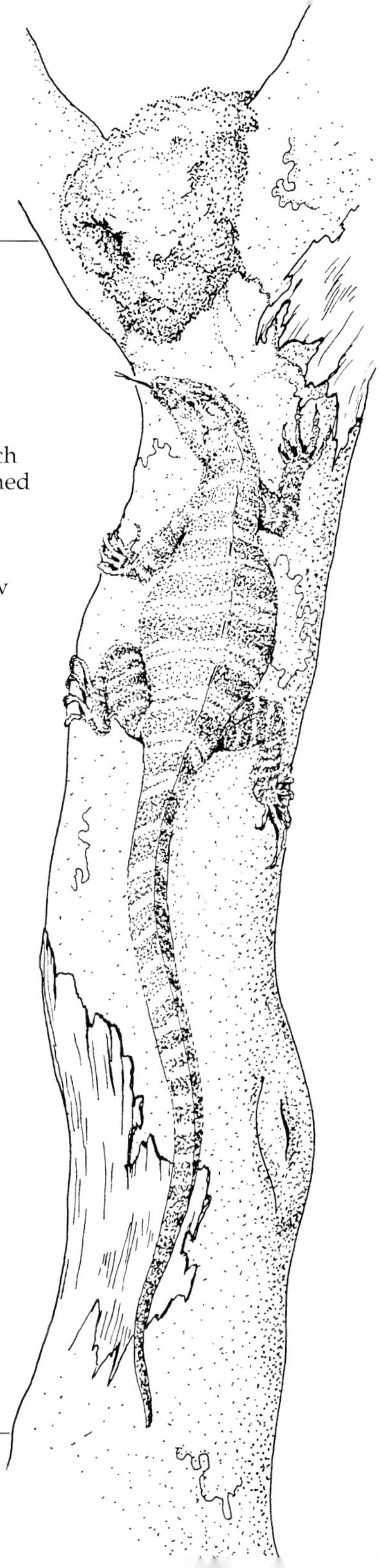

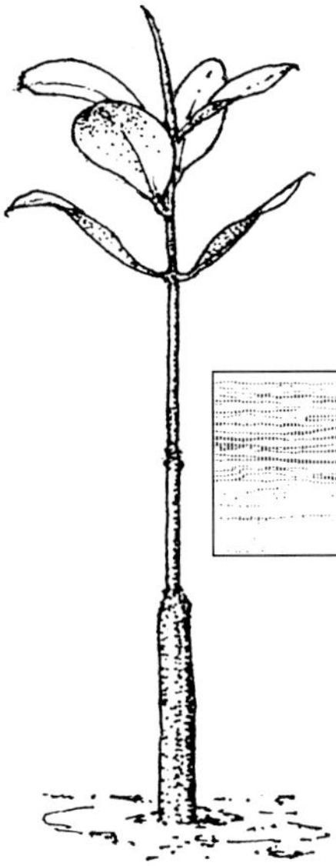

MANGROVES are trees living on the edge of two worlds. Where the sea and land meet, they are able to cope with their roots being covered by salt water at high tide, often exposed to air at low tide, and survive in water that is almost fresh after heavy flood rains. Their living conditions are subject to constant change. Water heights fluctuate daily, salinity levels and nutrient supplies alter, while tidal changes affect the oxygen supply in soil and water. They are indeed special trees to handle all these variations.

Seven different species grow along the Noosa River. You can see them in the 'Frying Pan' arm of the estuary, along Weyba Creek, R.343 (the reserve opposite the Australian Rules Football Ground) and Keyser Island, around Goat, Sheep and Pig Islands near Tewantin, Lake Doonella and around Cooroibah Creek. Some sections of shoreline around Lake Cooroibah and Lake Cootharaba have mangroves, as well as a narrow fringe along the river banks. The end of the tidal influence is seen with the last mangroves found around Kin Kin Creek's junction with the wetlands.

It is on R.343 that all the different species grow, and the change in mangrove vegetation zones can be easily seen. The first line of defence against the scouring action of creek currents and waves from passing speedboats is the Black Mangrove with its looping roots. Another name is the Small-stilted Mangrove. Those arching prop roots create a wonderful buffer zone, trapping organic matter and protecting the mud banks. By sending down roots from branches, and leap-frog roots as well, it creates a veritable maze that is very difficult to negotiate. This tree has large leaves and small white flowers.

Behind them, and growing in the mud which is covered twice daily with water, are the Goat's Horn, the Orange, the Spurred, and the White (often called Grey) Mangrove. The first three names refer to the shape of the fruit of each tree. The last refers to the trunk's colour. Each has a different root shape which helps to stabilise the flotsam and silt trapped by the outer trees.

Each also has a different system of gaining oxygen from the dense anaerobic mud, using pneumatophores above ground level either jutting up or bulging out from lower roots. Most easily recognised are those of the Grey Mangrove. It sends up numerous peg-like pneumatophores which have lenticles – porous areas able to absorb oxygen from the air – be they on roots or stems. All these trees living in salt or brackish water have another unusual trait. They can extract salt from the water and exude it through pores on both leaves and branches. Run your finger over a Goat's Horn Mangrove's leaf, for example, and taste the salt!

Each species has its own special way to ensure its seeds germinate and have a good start in life. The Black, Orange (or Large-leafed) and Spurred Mangroves have the unusual way of allowing the young plant to develop while still on the tree. The seedlings form first as swellings within the spike-edged flower, then into long hanging shoots. These grow thicker at the bottom end and drop off from Black Mangroves when they are about 10 cm or so. Some float away, bobbing vertically in the water at high tide, while others plop straight into the mud. Roots seem to shoot as soon as the seed touches the mud and the young plant grows quickly, consolidating its hold.

Lower in height and usually growing around the inland side of the group is the Spurred Mangrove. It becomes a small tree with rather yellowish erect leaves, buttress-like roots, and grows on the edge of very salty areas. Amongst the Spurred Mangroves on the landward zone more Grey Mangroves grow, plus Milky and River Mangroves. The Milky Mangrove exudes a white milky sap from any broken area. This is said to be painful or even blinding if splashed in the eyes. Its fruit is a three-hooked capsule which bursts open violently to free the seeds. The River Mangrove likes the landward zone where seepage occurs, and also the seaward fringe of tidal creeks, for it has a high range of salinity tolerance. Clusters of small scented flowers have nectar attractive to bees, and are important honey suppliers for apiarists.

As the land becomes less saline, more elevated, and has more oxygen and nutrieents, then other trees are able to grow. Swamp Oak, Swamp Box, tall Paperbarks or Melaleucas appear, and finally Forest Red Gums and Moreton Bay Ash.

Few people appreciate the complexity of life in mangroves, and the completely specialised life of trees, birds, fish and invertebrates that depend on this area for their life. How important that large areas should remain, for they also provide a nursery for our local fish and prawns. We must ensure protection for the surviving mangroves no matter what the pressure to nibble into or obliterate any part!

WHITE OR GREY MANGROVE TREE

This tree sends up a series of vertical peg-like pneumatophores from its roots. Projecting above the dense mud, they absorb necessary oxygen for the tree.

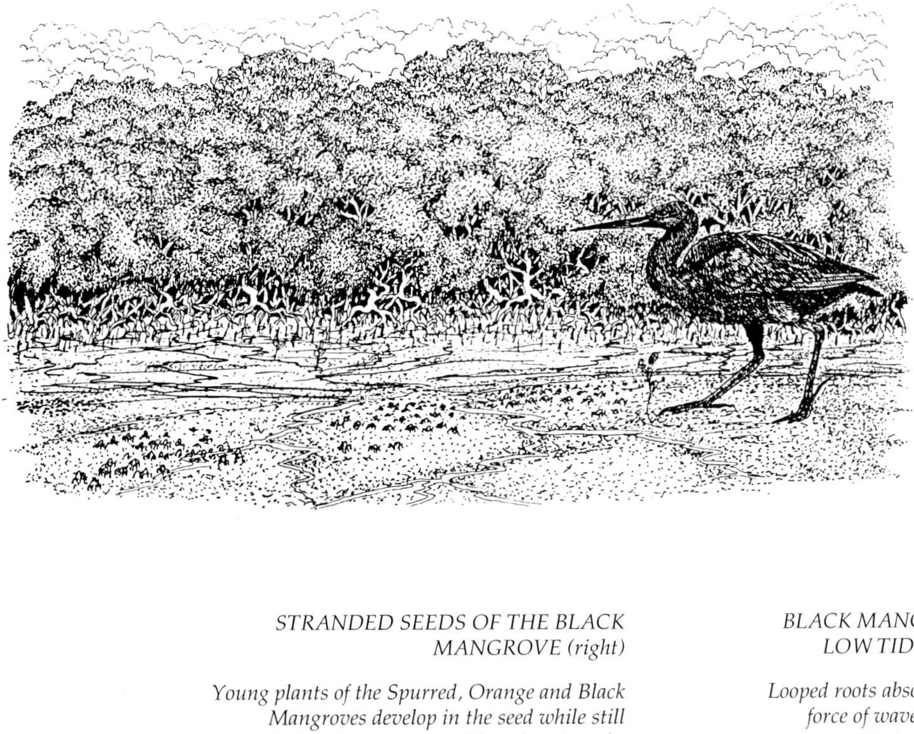

STRANDED SEEDS OF THE BLACK MANGROVE *(right)*

Young plants of the Spurred, Orange and Black Mangroves develop in the seed while still attached to the parent tree. They drop into the mud to take root, or float away. When one manages to maintain a vertical position at low tide it can commence growing.

BLACK MANGROVE AT LOW TIDE *(far right)*

Looped roots absorb the main force of wave action, and protect other species.

SECOND POINT, LITTLE COVE BEACH AT LOW TIDE

Littoral zones are revealed, while shells, crabs, anemones, cunjevois and even blue-ringed octopuses are there for beachcombers to observe.

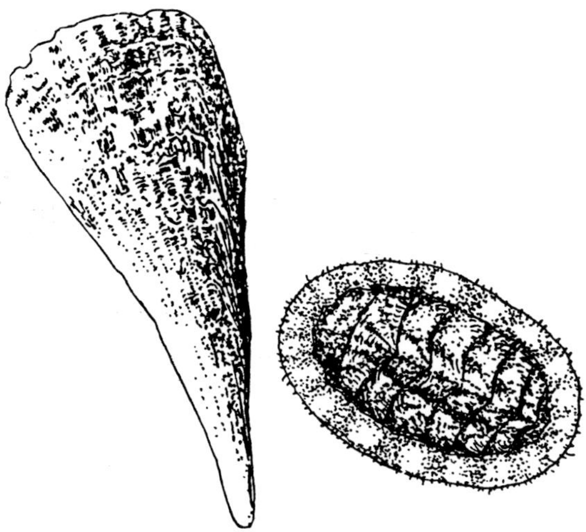

PANDANUS

Segmented, globe-shaped fruit of the Pandanus palm turn from green to an orange-red when ripe. Then it falls, breaking into small pieces.

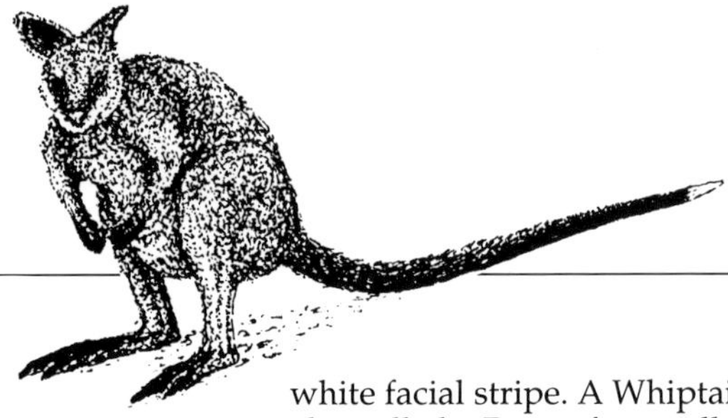

COULD THE Swamp Wallaby be considered common in the Noosa bush? Relatively speaking, yes, but it is a shy animal and usually seen only dead on the road or heading quickly into thick undergrowth. Neither of these is an appropriate time or place to pause and appreciate the heavy build and sombre colouring of this study marsupial.

If you are able to examine it closely, you will find its body fur to be thick, shaggy, and a dark brownish grey. Its belly fur by contrast is rufous to yellowish-orange. Its two distinct body colours give rise to its scientific name *Wallabia bicolor*. The muscular tail's body half is a dark grey while the end half is quite black. Here in Noosa there is a variation where some have a white tip. Look closely at the face and you will notice its short ears, and an indistinct grey facial stripe.

The largest of Noosa's wallabies is the Rufous or Red-necked, with some 107 cm of head and body plus 76 cm of tail. Next comes the Agile Wallaby with its prominent white hip stripe. It is slightly smaller, with a head and body some 94 cm long and a longer tail of 87 cm. The Whiptail has the same body size as the Agile, and its tail is just 5 cm shorter. However, its quick identification comes from a distinctive white facial stripe. A Whiptail is also called a Pretty-face wallaby, and when you see one you will understand why.

Next smallest is the Swamp Wallaby with its 84 cm long head and body, and 66 cm tail. As the name suggests, it likes to live in damp areas. You can find them in damp thick gullies, hopping through mangrove shallows, haunting the hillsides through open timber country or wet sclerophyll forest, or even in rainforests. However, they have also been observed in open heathland, and up on the Cooroy Range. Lantana and groundsel thickets offer lots of protection.

Swamp and Brush-tailed Wallabies are sometimes confused when being identified in rough hilly country. Both have similar colouring, but the first has a typical tapering wallaby tail while the second is long and cylindrical.

One seldom sees more than one Swamp Wallaby at a time. They dislike gathering into a mob, and are definitely 'loners'. However they sometimes feed together, and must do some courting. Apparently there is no definite breeding season, and females produce their single offspring at any time of the year.

Startle a Swamp Wallaby and it will take off very quickly, head down, tail straight out behind, bounding low to the ground rather than hopping in the accepted kangaroo style. Hands are not tucked in close to the body but held out. It does not stop after a short distance to sit upright and peer back at its disturber as a Whiptail does, but keeps on going, crashing further and further through the undergrowth.

The Swamp Wallaby is essentially a vegetarian, preferring to browse on shrubs and bushes rather than grass. Researchers have found that it likes to eat bracken fern, a plant which can be poisonous to cattle but apparently has no effect on this wallaby. Young pine seedlings, and agricultural crops that they can reach from shelter are also munched.

Both the Agile and Swamp Wallabies are unusual in having a broad pre-molar which remains in the jaw and is used for breaking off coarse food material.

Loss of habitat has certainly affected the numbers and distribution of this wallaby, but it has little to fear from shooters. It is too small, and its fur too coarse for commercial interest.

Hunting by dingoes and domestic dogs poses the most danger, and add to that the danger from road traffic, especially at night. Once dazzled by headlights the animal loses its sense of direction completely and is just as likely to jump into the car as away from it!

Although many have been killed by vehicles, Swamp Wallabies are not an endangered species. There are many in the Noosa bush, which suits them well.

IS THE Short-necked Tortoise really a turtle? Technically speaking the answer is yes, but because Australia has no purely land-dwelling tortoises, all our clawed, web-footed freshwater turtles are commonly called tortoises. For convenience sake the name turtle is confined to marine turtles with paddle shaped flippers.

Three common species live in Noosa's lakes, dams and streams. All are dependent on water for their food supply, ranging from aquatic animals, fish, crustaceans and molluscs to insects. Tortoises also eat some submerged aquatic plants including Water Hyacinth (both its flowers and bulbs), which although beautiful is a noxious weed.

The Short-necked Tortoise is particularly fond of freshwater mussels. Strong tortoise jaws can crush small shells, or snap off the tapering edge of larger ones. The mussels quickly die and relax, enabling the tortoise to prise open the shells.

They are reptiles completely adapted to freshwater living. They feed mainly in the early morning or late afternoon when the angle of the sun's rays does not cloud the water. At midday they emerge and lie basking in the sunlight on some projecting log or boulder, though often the only sign of their presence is a nose with two small nostrils above the water. Little black blobs about the size of your finger tip seem to keep disappearing from the placid surface, leaving only expanding disturbance rings to mark where they were.

How can one distinguish between male and female? Two simple rules may help. The male's tail is much longer than the female's tail, and she has a much deeper body.

There are two causes for a tortoise to leave its watery element and travel across the land on those short clawed legs. They are to lay eggs and to migrate. Of course only the female seeks a suitable soft river bank nesting site. It may be as high as 100 m up a steep river bank like the Mary River, or as far as 500 m away in swampy country. Unfortunately, the further the eggs are from water the more hazardous for the baby tortoises as they return for safety and food.

At the chosen site, whether sand, soil or clay, she digs a hole about 15 cm deep. Her cupped hind feet dig alternately. Then about twenty white, oval, leathery-shelled eggs are laid, carefully covered with earth, and padded flat by the back feet. She drops her body several times to consolidate the soil, finally

The Long and Short of Tortoises

raking across the surface to disguise the site.

Incubation times vary, but on average three to four months elapse before the youngsters emerge.

Another common tortoise living in the Noosa Shire is the Long-necked. Like all Australian tortoises, it retracts its neck sideways under the carapace, and there is a lot to tuck in. When stretched forward the thin neck is almost equal to its shell in length!

Both tortoises migrate, but the reasons are not fully understood. Large numbers have been known to leave drying waterholes and travel in one direction, presumably in search of more permanent water. Quite understandable, yet the same mass migrations have been observed in the same direction following good rain!

Changeability of weather, extremes of cold or heat, and drying up of waterholes are all coped with by our tortoises in special ways. No hibernation occurs here as it is too warm, but they do go into a state of torpor called aestivation.

The tortoise, after drinking as much as it can, leaves the drying pool to seek a damp place. Here it buries itself just below ground level. It could be under a fallen log, amongst foliage, in a cool rocky crevice, or in a suitable hole. Some go to the bottom of a muddy waterhole.

As long as the drought is not too long they emerge when it rains to continue an active life. Resting on half-submerged logs, or only poking two nostrils above the surface, they are a welcome part of the Noosa scene.

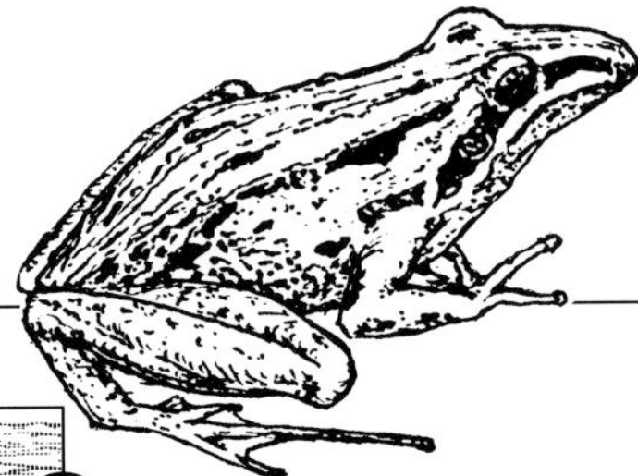

SUMMER RAINS bring such a variety of frog calls from any dam, stream or swamp at night. Some frogs don't wait for nightfall. The Dwarf Tree Frog, for example, calls its 'wreeek-pip-pip' from shrubbery by day if it is still raining. The large Green Tree Frog seems able to foretell rain by a couple of days and calls out its 'crawk crawk' from house down-pipes and drains.

But if you listen to the night chorus coming from a nearby dam, what a variety of calls there are.

The introduced Cane Toad's call may dominate with its rapid 'put-put-put-put' noise. Listen past it to the plops, bonks, trills, quacks and pulsing whistle sounds of others. There are thirty-one different species of frogs found in this area, some living in mountain streams, some in dryer areas under damp stones, some in marshy paddocks, while others like reed swamps and damp heath.

One interesting frog likes coastal streams and dams with good grass cover all around. It is the Rocket Frog, a winner in any frog jumping competition. On a wet night your car headlights have probably shown you one jumping across a country road. A swamp or dam would be nearby.

It has a light brown (almost bronze) streamlined body with its colouring broken by black line marks. Down its back run four rows of lumpy warts arranged in two series with a paler area between.

A strong black line runs through the eye from the very 'Roman' nose to behind the forearm. A few white cross lines intersect it, including the light edge to the large tympanum or ear disc. The eyes protrude right out of the head – wonderful golden-brown eyes with a lighter horizontal strip above the pupil.

The nose is impressive, sharply pointed, with a distinctive, down-turned profile. No wonder the frog has been given the scientific name *Litoria nasuta*, nasuta meaning 'the nose'.

But the spectacular leaping ability contained in its extra long legs has given this amphibian its most apt common name, Rocket. Sometimes it can span a narrow bitumen road with only two or three leaps, and six metre jumps have been recorded. Look for it in the grass afterwards and it will be very hard to see. It has the habit of remaining motionless once it reaches cover. Being between four and five centimetres long, it is a medium-sized member of the tree frog genus. All members of this group have suction pads on the tips of their fingers and toes.

What a Jumper!

Because the Rocket Frog spends so much time on the ground its discs have degenerated. They are so small as almost to escape notice. However, one does notice the great length of the hind toes, because the fourth toe extends out by itself from the foot. Both its disc and the end two phalanges (the bony toe sections) are free from webbing connecting them to the toes on either side, as also are the disc and half a phalange of the fifth toe. Thus when at rest this long thin toe protrudes out some distance from the long but closely folded hind leg. The fingers have no webbing at all.

How can the call be described? Densey Clyne calls it an excited nasal 'yapping', and several calling together have a gobbling affect like the word 'double-u' repeated faster and faster. Two other experts – John Barker and Gordon Grigg – describe the call as starting irregularly and building up to a rapid duck-like quacking.

Whatever words we humans use to describe it, female Rocket Frogs find it irresistible. Although to human ears two other local frogs closely related to the Rocket share his call, each frog attracts only its own species.

However interesting or different the call may be, it is the sight of this frog's prodigious leap that remains in our memory.

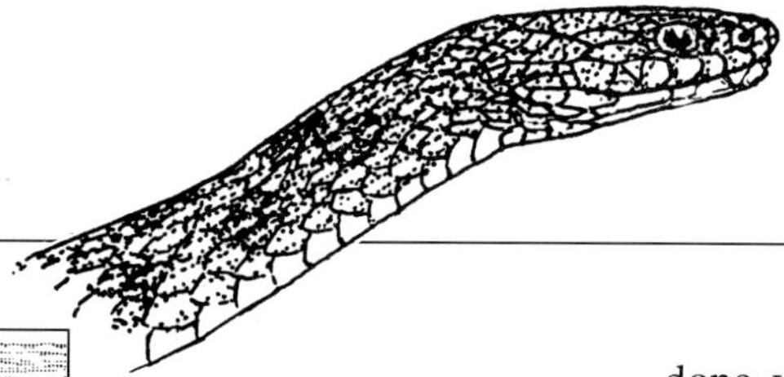

NOT EVERY Australian snake is a danger to man and beast - far from it! Although it is true that some like Death Adders and Taipans are very 'nasty' and should be treated with caution and a great deal of respect, others are quite harmless. The latter have solid teeth and no poison glands, so if you annoy them and they defend themselves, a row of teeth marks is all you will receive. Pythons and Green Tree Snakes are included in the group, as well as a fairly common snake around Noosa's dams and streams, the Freshwater Snake.

Unfortunately this harmless snake is very similar in outward appearance to the dangerous Rough Scaled Snake. Both have scales with sharp ridges, something you can see quite easily. Both are found in similar areas (Nambour district is one), and both are dorsally a dull grey-green with darker bands. The Tiger Snake has similar markings, but though these are common in the south, there are few found north of the border ranges. It also has no ridged scales.

Colour is usually an unreliable guide to identification, as so much variation has been recorded. A scale count is necessary for correct identification. The trouble is, the snake needs to be dead before that is done, which defeats the idea of sparing the life of harmless freshwater snakes!

If you do come across a dead one, here are the figures for the two species.

	FRESH-WATER	ROUGH-SCALED
Scales across body	15	23
Ventral scales	140-160	165-182
Anal scale	Paired	Single
Subcaudal scales (under the tail)	50-80 Paired	50-54 Single

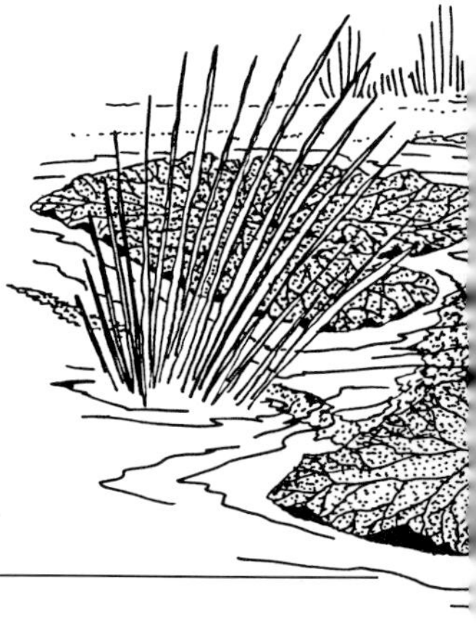

Serpent in the Reeds

As the name implies, the Freshwater Snake is a strong swimmer. However it appears to swim no more than, say, the Red-bellied Black. Purely aquatic snakes like the File Snakes of north Queensland have nostrils high on their snout, but not so the Freshwater. It hangs around water because frogs are its main bill of fare. Where food lives, so too will snakes. When not out at night hunting its tucker, you may find it curled up in a hollow log nearby, under a sheet of bark, or under galvanised iron sheets and other man-made items lying on the ground. The Freshwater Snake's leathery-shelled eggs may also be found in these hiding places in summer.

If you do capture and handle one it rarely attempts to bite. Instead it may drop its tail and give off an unpleasant smell as it tries to escape. The Rough-scaled Snake (which is often called the Clarence River Snake) on the other hand can be particularly aggressive. As these two are so easily confused when seen out in the field, it is wisest to leave both well alone. All too often has the harmless Freshwater been killed through a confusion of identity. Perhaps this last difference will help save its life. The Freshwater grows to about 1.2 metres long, the Rough-Scaled to 0.65 metres. So if you come across a snake near water with keeled scales and it is well over 0.65 metres, do not kill it. It is harmless!

WHAT A strange-looking bird is the Spoonbill. Have you ever been close to one as it walks around feeding in the Noosa River shallows? High tide finds these birds perched in nearby mangroves, but at low tide when the water is less than 40 cm deep they commence feeding, swinging their unusual bills from side to side, searching for worms and molluscs in the mud as they walk along. Perhaps you have seen them in shallow dams and streams. Lake Macdonald's shallows are visited frequently.

As a small flock flies to a new feeding ground, watch these large white birds with long black legs trailing and long wide black bill held out straight ahead. The bill looks too heavy to be held out horizontally, but like all large bills it is almost as light as sponge rubber, being honeycombed with air chambers.

In flight a Spoonbill looks rather like a White Ibis, but the ibis bill, though also long, is thin and downcurved. Another large white bird is the Egret, but it always tucks its head back to its shoulders when flying, and it has a straight pointed bill. No cause for confusion there.

I mention Royal Spoonbills for they are the species most often seen near the coast. The other species, the Yellow-billed, is more common around inland streams and lakes and has greater numbers. Though the Royal is slightly smaller, both birds are similar in most aspects. For example, both have slender, widely spread toes that keep them from sinking into soft mud, and long necks to counterbalance long legs.

Whether black or yellow, it is the bill that fascinates. The wide spatular ends are very sensitive and adapted to finding crustaceans and small fish as it is swept from side to side. If a prawn is caught it is first crushed by horny knobs at the base of the bill before being swallowed with a backward flick of the head.

Unusual though its shape may be, the bill is capable of doing the same work as other bills. It dresses plumage, communicates with a 'chew' call and much clapping together of mandibles, it holds and weaves sticks and reeds for nest building, ministers to the needs of the young, manipulates food, and even defends its owner!

Spoonbills are normally shy birds, but at nesting time when nuptial plumes grow on their head and neck they often form a loose colony with Ibis, Herons and sometimes Cormorants. Their nest is a platform of broken down reeds and twigs usually built in a bush or tree above

Regal Swinger

water. There up to four eggs are laid, and both birds share nest building, incubation, and later feeding the chicks.

The parent bird feeds as usual in the shallows, swallowing all the tucker, then it returns to the nest to regurgitate the half-digested worms, fish and so on into the chick's mouth. However, the method they use is unusual. The parent grabs hold of the hatchling's beak crosswise in its bill and shovels the food down into the young bird with its tongue. As gastric juices go down too, the youngster finds the food very easy to digest!

Should another bird come too close, the Spoonbill will raise those nuptial nape feathers like a fan, and act aggressively towards it. The intruder will see the bright red forehead skin where the feathers commence, its bright red eyes and yellow eyebrow patches, plus the 'shot silk' slate blue patterning on the bill itself. Humans seldom get that close.

Royal Spoonbills are frequent visitors to Noosa. Watch for them with their swinging bills.

LAKE POONA,
COOLOOLA NATIONAL PARK

A pristine gem of a 'white water' perched dune lake, it is slowly becoming cloudy with organic matter and algae caused through pollution by its many human visitors.

RIVERSIDE REEDS

These provide a natural protection from the wake of passing boats, but they grow only along sections of the banks. Erosion is a constant threat.

RED SPATULA SUNDEWS

Their special leaves glisten with sticky secretions, ready for small insects to wander on to, and become trapped. These plants bordered a freshwater lake.

REFLECTIONS ON THE UPPER NOOSA RIVER

Tree-lined, narrow, winding and deep, sections of the river are usually so protected from wind that they present a glassy surface. Reflections are intensified by the dark tea colour of its waters.

EACH YEAR Noosa celebrates the Festival of the Waters, focusing attention on the lovely river and its shallow expansive lakes. The river's beauty is an integral part of the Noosa scene, and needs protection from pollution and detrimental exploitation.

People enjoy the river on sailing dinghies, windsurfers, yachts, houseboats, launches, rowing boats, canoes and inflatables, sharing with pelicans, terns, cormorants and herons the living water of our river system. It is indeed still a living river, both above and below the water line. We all benefit from that, whether through swimming, boating, fishing, or merely as a spectator on its shore.

From thinking of the many kinds of human boatmen enjoying its saltwater reaches, it is but a step to the Water-boatmen, Backswimmers and Whirligig Beetles of its placid freshwater upper streams. All these insects live in water, though of course they are found only above tidal influence. There, with the correct balance of temperature, sunlight, oxygen in solution, food, and living space these specialised underwater insects thrive.

Oxygen is essential for gill-breathers like dragonfly nymphs, and living plant life provides it in great quantities by releasing it into the water though photosynthesis. These plants are also a food store for the vegetarian Water-boatmen. Plankton which grows well in the shallows is food for tiny protozoa and crustacea that hide amongst the plants, and these in turn become tucker for Diving-beetles and Backswimmers. The food chain continues as in turn they become food for dragonfly nymphs, frogs fish, birds, and so it goes on. Anyone having the time and opportunity to watch carefully would be amazed at the activity that goes on above and below water in quiet streams.

Water beetles are specialised insects living submerged yet not having totally lost their dependence on surface oxygen. They periodically rise to the surface to breathe, protruding the end of their abdomen out of the water and replacing air in the trachea system. Hidden underneath the hard elytra or top wings, the surface of a water beetle's abdomen is covered with fine hairs that form a thick pile. By raising its elytra air is trapped there and retained as a bubble after each surface visit. This happens only in the adult stage. The long thin larva has to rely on periodically rising to the surface to take in new oxygen through protruded rear spiracles.

Water is their Element

Water Beetles use their flying wings at night to travel long distances, seeking out new food and living ponds. Night lights often attract them to houses, when you can readily examine their specialised swimming legs. Your swimming pool may receive a sudden visit, but rest assured that after they have cleaned up any food around, they will go.

A Backswimmer is unusual. Here is a small aquatic bug which propels itself through the water with its keeled back downwards, using its long hind legs as oars. These are thickly feathered with hairs and are most effective. The male makes an unusual chirping sound to attract females. It commences as a series of squeaky notes that increase in tempo and pitch until it becomes a shrill metallic chirrup, something like a cicada's call. But instead of a cicada's abrupt ending, this love song fades gently away.

Of all water beetles, the Whirligig appears the most active, attracting most attention. Their sleek black oval bodies whirl around each other on the surface, sending out ripples, and darting everywhere. They like company, and often behave like 'dodgem' boats. Being surface movers they have specialised eyes. The top half is for viewing above, the lower half for seeing underwater. Their middle and hind legs are much smaller than the front pair, but are wide, flattened, thickly fringed with hairs, and paddle madly along underneath. The front pincer-like pair hold prey. Water beetles have small 'hidden feelers' that fit into a groove for streamlining, and mandibles that suck out the juices from prey.

If a Whirligig dives under the surface it takes with it a bubble of air at the abdomen's tip, like an aqualung diver's oxygen. However it soon returns to continue its crazy whirling.

Whether you enjoy the river and imitate these insects as a diver, swimmer or by just splashing around, consider other users both seen and unseen. Whether you move on, or under, or merely sit beside the water, value its natural unspoilt beauty and help maintain it. Future generations would like to enjoy it too!

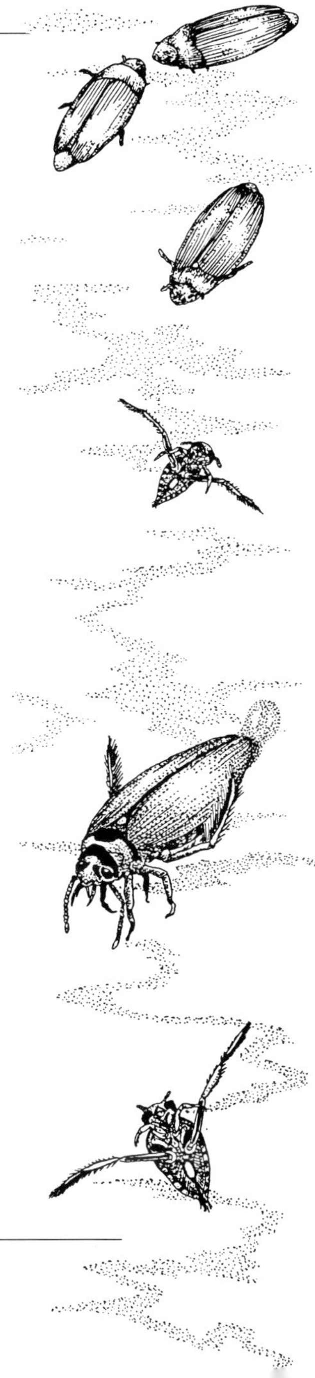

AROUND THE EDGE of Noosa's freshwater lakes and dams you will find growing a most interesting plant, the Sundew. Not one or two, but hundreds could be crowded around the margin of a swamp, lake, or merely in a wet depression. Although denizens of low-lying ground like the Bladderworts they are often found growing at higher levels in roadside drains, or where impervious rock or clay holds moisture near the surface. All they usually ask for is wet feet and sunshine.

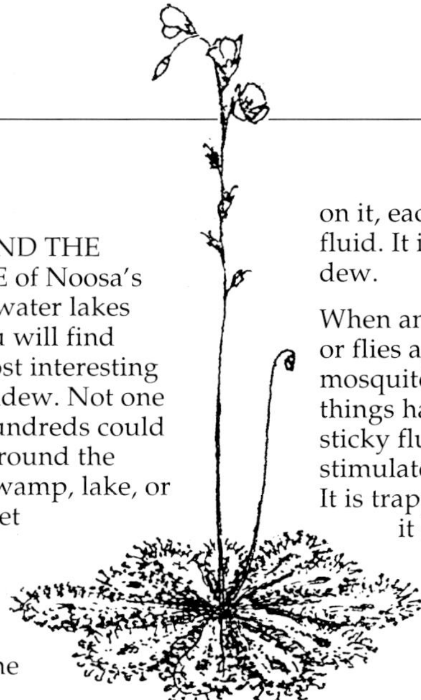

Three different species of the genus *Drosera* are commonly found in Cooloola and coastal areas, and they are all known by the popular name of 'Sundew'. All are small and have what appear to be dew drops clinging to their leaves, glistening in the sun.

Most widespread here is the Red Spatula Sundew. It has a rosette of spoon-shaped leaves snuggling close to the ground, the whole plant covering no more than 2.5 cm. Each leaf has numerous glandular hairs on it, each covered with a sticky fluid. It is the fluid that glistens like dew.

When an insect wanders onto a leaf or flies against it (as an ant or mosquito might do) it finds strange things happen. It is held by the sticky fluid while its struggling stimulates the hairs to bend inward. It is trapped. Struggling only forces it closer to the leaf! Then the plant secretes fluid enzymes which digest the insect. The nutrients flow back through the leaf to nourish the whole plant. Although Sundews are like other vegetation and obtain energy from photosynthesis, they appear to need extra nitrogen and obtain it from trapped insects. Nitrogen is deficient in the coastal region's acid soils.

One often sees another sundew with different leaves, though they perform the same insect-trapping job. No common name means it is known as *Drosera peltata*. When very young it grows short broad rosulate leaves on the ground, but these soon disappear when the thin stem rises (up to 40 cm tall and sometimes branched).

However the plant is still able to capture insects for dinner, as new small leaves grow fairly evenly

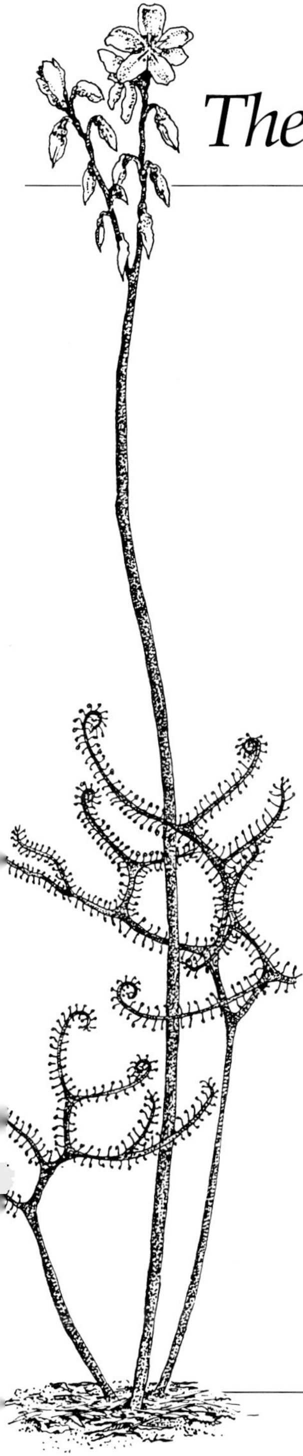

These Plants Eat Flesh!

spaced up the thin pliable stem. They are complete with sticky hairs and liquid ready to carry out their function. If you do see it flowering at the top the flowers will be isolated, four-petalled, 5 to 6 mm long, and either white or pink.

Tallest of the three common local plants is the Forked Sundew. Leaves of this plant are not found in a cluster on the ground, or dotted up the stem, but grow on long stalks that can range from 5 cm to 60 cm long. The leaves are long and thin, dividing into forks several times, each tip being curled. Their short sticky hairs grow all around and along them.

This plant also has the largest of the flowers, each of the five delicate white petals being about 1 to 1.2 cm long. They can bloom as high as half a metre from the ground, though usually with the support of surrounding vegetation.

Many people enjoy growing sundews in their own home garden, but they do best not in beds but in pots. So long as the pot is kept damp by standing it in a saucer of water, and is kept in moderately strong light it should do well. Sundews transplant easily, and can be grown by seed, or bought from a native plant nursery.

But whichever sundew you grow, you will be intrigued by its bizarre method of catching insects, and delighted at the delicacy of its simple flowers.

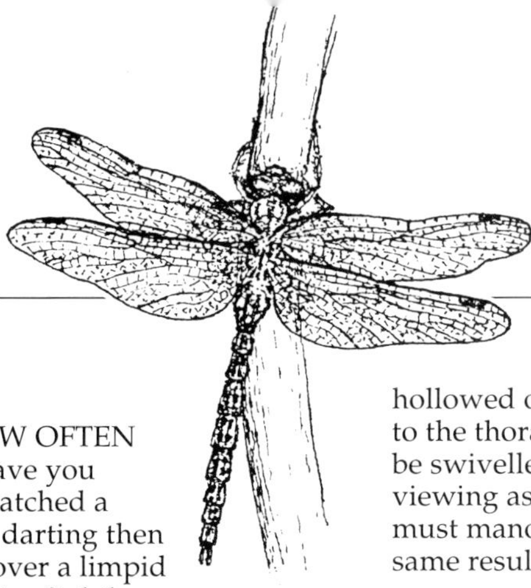

HOW OFTEN have you watched a dragonfly darting then hovering over a limpid pond, resting lightly on a reed or lily pad, then darting off again. Its flash of blue, or red, or green, or brown is as much a part of a summer's day as a cicada chorus, or chirruping crickets.

A dragonfly is such an interesting insect, spending most of its life as a ravenous nymph underwater, then a few weeks as a winged insect searching for other insects to eat. In flight, it reminds me so much of the helicopter which makes regular beach patrols on summer weekends. That can rise vertically, hover, and progress laterally just like a dragonfly. There are other similarities, for dragonflies have a heavy head and thorax compared to their long thin abdomen, just as the helicopter has a large motor and cabin in front of a long thin tail.

The dragonfly's thorax houses extremely strong muscles. Its huge bulging multiple eyes are so large they sometimes meet in the middle, and really dominate its head. Thus it can see in all directions just as the helicopter's large perspex bubble gives it maximum visibility. In one way they are dissimilar, for the dragonfly has the rear of its head hollowed out and loosely attached to the thorax. Its huge eyes can then be swivelled around for even better viewing as it searches for prey. Man must manoeuvre his craft for the same result.

Dragonflies have four wings as opposed to the 'copter's overhead rotor blades. However helicopters have smaller blades rotating vertically in the tail to prevent the machine from swinging around from the force of the overhead rotor. Both the wings and the lifting blades are long, slender, and braced so as to be strong yet pliant.

Alone of all insect wings, the dragonfly's fore and aft pairs beat quite independently. The front ones rise as the rear ones fall, thus enabling it to remain quite stationary, suspended in the air. This is the way the wings worked back in Palaeozoic times, and it is still considered the deftest of flying insects.

The wings are rather special in another way, for they are not like modified limbs but are part of the insect's back, being basic structural outgrowths of its external skeleton. Each is delicately strutted with passages for air and blood.

Look closely at a wing if you have the chance. The hind one is usually broader than the forewing, again something different in the insect world. There are no covering scales

Nature's Helicopter

(like moth wing scales) but a clear stiff membrane which often has an iridescent metallic sheen and occasional thickening into bars and spots.

A wing's forward edge requires extra strength and has a concentration of veins. Behind that (especially in the dragonfly) there are delicate reticulations and struts which can contain as many as three thousand cells. A strong cross vein comes about halfway along the leading edge, and the identification of a particular species depends on the size and positioning of its veins.

Earliest fossilised insects are revealed to us as having four separate functional wings just as dragonflies have today. Most modern insects work their fore and aft pairs as single flight surfaces. Moths have special hooks joining their fore and hind wings together, and butterflies overlap their wings to function as a single surface. Flies on the other hand have hind wings reduced to small stump-like appendages, which act as stabilizers.

None of these insects has the dragonfly's capability of long sustained flights, soaring, hovering, and darting. An amazing speed of 100 kph has been recorded for one species.

Our Noosa insects are like all the others, masters of their medium. They need no instruction on lift, drag, wind velocity, and how to solve the complicated equations involving weight, wing area, rate of vibration, air density, speed of flight and so on. They know instinctively how to twist and turn in flight, changing direction so suddenly that the human eye cannot follow them. The helicopter has not managed that art, though it is on the way.

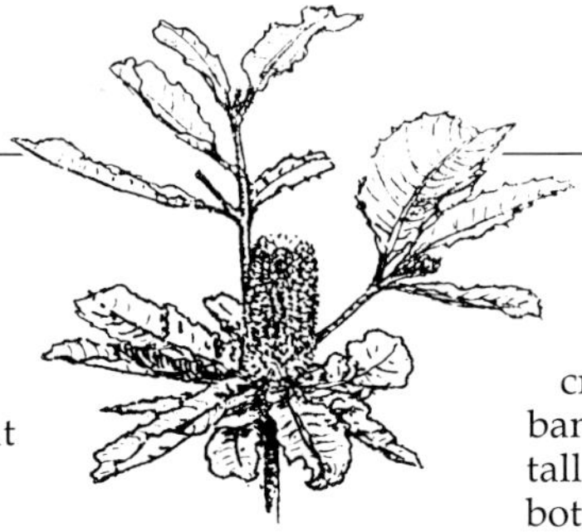

THERE ARE few flowers in nature that are a rich green in colour, but we have one at Noosa. It flowers in autumn, and is well worth a pause and close examination.

Look over any damp heath country and you will see the large, glossy, stiff leaves of *Banksia robur* amongst and above the vegetation. It is known as the Swamp Banksia but is not confined just to swamps.

Even in sandy forest areas it grows well, as long as the soil is damp and it is in a sheltered, shady position. Within the forest conditions of a melaleuca swamp this woody shrub can grow as high as three metres, while out on an open heath where there is so much sunlight it reaches only about a metre.

As you look across low heathland, notice where these large leathery leaves stand up above the rest of the vegetation. You may be sure that particular area is very damp under-foot.

Usually the plant grows to about two metres amongst Hakeas, Wedding Bushes and other trees. Look along the Noosa River banks, particularly in Cooloola National Park, and you will see some fine specimens.

Most banksia flower spikes are cream, but when these banksia buds first form as tall candles they are a rich bottle green colour. Find an older spike and you will find it has changed to a dark bluish-green. The perianth limbs have changed, and as the flowers proceed to open they turn to yet another colour, a yellow-green. Then look closely at the flowers. You will see that the tiny stigmas at the ends of the spreading styles are tipped with black. All the flowers on the spike open from the base first.

Slowly the colour changes yet again, this time from age. The flower now becomes a rusty, brownish red. Old flowers remain on the shrub for some time, so one can admire all stages of the colour change on the one bush.

This rusty red is similar to the colouring of its new leaves which sprout from the tips of its woody branches. Fine closely matted woolly hairs also cover the new leaves and stems, hairs that continue on the underside of the stiffening leaves as they grow to maturity. Then they become up to 30 cm long by 10 cm wide, with irregular shallow serrations along the edges. Even the tip has a sharp point, yet most leaves show damage from insects. Notice how the orange midrib and bright yellow parallel veins stand out against the dark glossy green.

On mature stems even the bark has a downy grey covering, though the soft, dense hairs of the Swamp Banksia's branches are more striking. Its specific name of *robur* (which means an oak tree) refers to the firm, oak-like leaves as well as to its strength and firmness.

Swamp Banksias are certainly tough shrubs. Even after a bushfire has swept over heathland, apparently killing all in its path, new growth will soon shoot up from the underground lignotuber. Before long those large leaves will provide shade, and seeds from other heath plants will germinate in the ashes. So continues the revegetation of damaged areas.

I am told you can propagate these shrubs from seeds, but make sure you raise them in a sunny, damp situation, preferably with rather acid soils such as we have here near the Noosa coast.

As you move around our coastal bushland, or laze on board a craft as it slowly cruises our river, remember to watch out for this rather special shrub with both green and rust coloured flowers. It is worth a second look.

WHERE CAN YOU see thousands of water-lilies flowering? At Lake Macdonald in springtime, with some flowers to be seen in bloom throughout the year. Many shallow inlets are filled from bank to bank with floating leaves of the Blue Water-lily, which prefers water about 50 cm to 75 cm deep, and lots of sun. It is a purely aquatic plant, growing with its roots and creeping rhizome deep in the mud.

Some native species have pink flowers, others red, but *Nymphaea gigantea* (which can be found from New South Wales right up to Northern Australia) can have either blue, white, or a pale pink flower. Blue is the most common.

At several vantage points you can leave your car and drink in the beauty of the flowers and nearby bird life. There are always birds on Lake Macdonald. Coots move though the leaves, each white face shield standing out against a slate-black body. Dusky Moorhens are nearby, each short upturned tail showing two thin white vertical bands beneath. Black Ducks cruise slowly along open passages, and White-eyed Ducks dip down for water weeds and aquatic insects.

Further out in the open water a pair of Black Swans glide along, possibly the same pair that raised a brood of cygnets on a large reed mound nest built on a mudbank. Little Grebes bob on the water, disappearing with a flip, and popping up again in unexpected places.

Decorating the bare branches of trees killed by the rising dam water are both Black and Pied Cormorants, wings outstretched to dry after some good fishing. Golden and Spangled Perch, Mary River Cod, Freshwater Catfish, and Eels are all found in the water, besides all the tiny non-commercial fish that Little Egrets enjoy chasing.

On a 'drowned' telephone pole a Darter (or Snake Bird) hangs his wings out also, and rattles his cackling call. Often a small flock of Marsh (or Whiskered) Terns fly lightly above the water. Suddenly one will wheel to dive in after a small fish swimming too close to the surface for its safety. Some of these birds wear dark breeding plumage in spring, but their nesting site is unknown.

Eastern (or Purple) Swamphens, each with a red bill, iridescent blue body and white under-tail feathers, join ever-alert Masked Plovers in keeping watch from the bank. Dainty Black-winged Stilts walk purposefully in the shallows with their 'shocking pink' long legs. Nearby may be small Waders, here

Lilytrotters and their Neighbours

from the Northern Hemisphere for the summer season.

But of all the birds enjoying the dam, its food supply and its water-lilies, perhaps the most interesting is the Jacana. You may know it as a Lilytrotter or a Christ Bird for it walks on water as it moves across the leaves. Water-lily plants are essential for Jacanas.

This small dainty bird has a body length of 20 to 24 cm, long thin legs, and exceptionally long toes. Both middle and hind toes are 7.5 cm in length, and with the others only slightly shorter its weight is distributed over 200 sq cm of water-lily pads when searching for food. Seeds, insects and aquatic plants are all eaten.

Flying involves a great effort to get airborne. Running over the lily leaves with much wing flapping is rewarded by flight, and it is then the long toes can be seen trailing behind the small body.

The Jacana's bright red frontal crest changes to yellow if it becomes disturbed or agitated. With a dull brown, black and white body it is usually the bright red crest one first notices as it moves about in the green maze of lily leaves.

If you have good binoculars you may find their springtime nest amongst the leaves. It is a low platform of water weeds and other vegetation on which the two or three eggs are laid, well camouflaged with wavy brown stripes.

Jacanas are very protective parents. Should the water level suddenly rise, or they need to abandon the nest for any reason, Jacanas carry eggs and even young chicks under their wings to transfer them to a new nest and safety. Few birds do that!

So many birds are found on Lake Macdonald at different times, from Pelicans, Magpie Geese, White-faced Herons, and Egrets to tiny Reed-warblers. It is well worth a pause to admire those blue water-lilies, and who knows, you might see a Jacana as well!

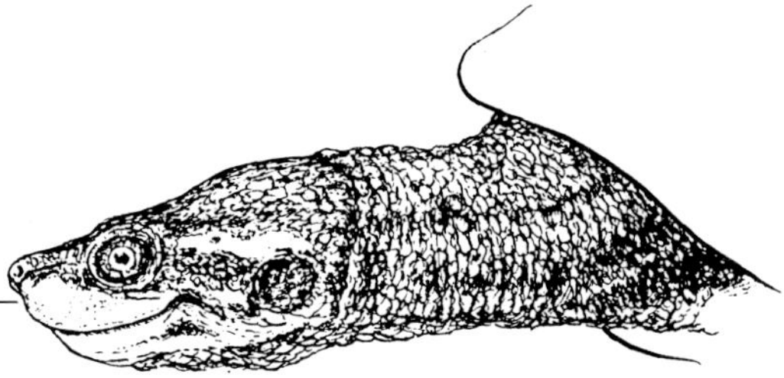

COELENTERATA
(Animals without backbones)

Portuguese Man-o'-war *Physalia physalis*
By-the-wind Sailor *Velella lata*

CRUSTACEA (Crabs etc)

Acorn Barnacles *Tetraclita rosea*
Sand Bubbler Crab *Scopimera inflata*
Soldier Crab *Mictyris longicarpus*
Ghost Crab *Ocypode spp.*
Hermit Crab *Fam. Paguridae*

MOLLUSCA (Shells etc)

Violet Snail *Janthina violacea*
Eugarie or Pipi *Plebidonax deltoides*
Hercules Club or Mud Whelk *Pyrazus ebeninus*
Warty Ear Shell *Haliotis ruber*

TUNICATA (Basic backbone animal)

Sea Squirt or Cunjevoi *Pyura stolonifera*

FRESHWATER FISH

Golden Perch or Yellow Belly *Macquaria ambiguus*
Spangled Grunter *Leiopotherapor unicolor*
Freshwater Catfish *Tandanus tandanus*
Freshwater Eel *Anguilla reinhardti*

SALTWATER FISH

Black Bream *Girella tricuspidata*
Silver or Yellow-finned Bream *Acanthopagrus australis*
Tarwhine *Rhabdosargus sarba*
Sand Flathead *Platycephalus arenarius*
Mud Flathead *Platycephalus fuscus*
Jewfish or Mulloway *Argyrosomus hololepidotus*
Sea Mullet *Mugil cephalus*
Tailor *Pomatomus saltatrix*
Golden Trevally *Gnathanodon speciosus*
Sand or Summer Whiting *Sillago ciliata*
Trumpeter Whiting *Sillago maculata*

AMPHIBIANS

Green Tree Frog *Litoria caerulea*
Rocket Frog *Litoria nasuta*
Dwarf Tree Frog *Litoria fallax*

REPTILES

TURTLES

Green Turtle *Chelonia mydas*
Loggerhead Turtle *Caretta caretta*

TORTOISES

Krefft's Short-necked Tortoise *Emydura krefftii*
Brisbane Short-necked Tortoise *Emydura signata*
Snake-necked Tortoise *Chelodina longicollis*

SEA-SNAKES

Elegant Banded Sea-snake *Hydrophis elegans*
Yellow-bellied Sea-snake *Pelamis platurus*

LAND SNAKES

Death Adder *Acanthophis antarcticus*
Taipan *Oxyuranus scutellatus*
Tiger Snake *Notechis scutatus scutatus*
Red-bellied Black Snake *Pseudechis porphyriacus*
Rough-scaled Snake *Tropidechis carinatus*
Freshwater Snake *Styporhynchus mairii*
Green Tree Snake *Dendrelaphis punctulata*

LIZARDS

Tree Goanna *Varanus varius*

INSECTS

BUTTERFLIES

Common Grass Yellow *Eurema hecabe phoebus*
Macleays Grass Yellow *Eurema herla*

Scientific Names of Species Referred to in Text

No-brand Grass Yellow	*Eurema brigitta australis*
Small Grass Yellow	*Eurema smilax*

BEES

European (Commercial) Bee	*Apis mellifera*
Native Social Bee	*Trigona sp.*
Solitary Bee	*Fam. Colletidae*

BEETLES AND BUGS

Assassin Bug	*Fam. Reduviidae*
Backswimmer Bug	*Fam. Notonectidae*
Water Strider Beetle	*Fam. Gerridae*
Diving Water Beetle	*Fam. Dytiscidae*
Whirligig Beetle	*Fam. Gyrinidae*
Water-boatman	*Fam. Corixidae*

OTHERS

Sand Wasp	*Bembix spp.*
Tree Termites	*Fam. Termitidae*

SPIDERS

Wolf Spider	*Lycosa spp.*

BIRDS

NON-PASSERINE (NON-PERCHING)

Australasian Grebe	*Tachybaptus novaehollandiae*
Australian Pelican	*Pelecanus conspicillatus*
Australasian Gannet	*Morus serrator*
Darter	*Anhinga melanogaster*
Black Cormorant	*Phalacrocorax carbo*
Pied Cormorant	*Phalacrocorax varius*
Little Pied Cormorant	*Phalacrocorax melanoleucos*
White-faced Heron	*Ardea novaehollandiae*
Mangrove Heron	*Butorides striatus*
Royal Spoonbill	*Platalea regia*
Yellow-billed Spoonbill	*Platalea flavipes*
Magpie Goose	*Anseranas semipalmata*
Black Swan	*Cygnus atratus*
Black Duck	*Anas superciliosa*
White-eyed Duck or Hardhead	*Aythya australis*
Brahminy Kite	*Haliastur indus*
Dusky Moorhen	*Gallinula tenebrosa*
Eastern Swamphen	*Porphyrio porphyrio*
Eurasian Coot	*Fulica atra*
Comb-crested Jacana	*Irediparra gallinacea*
Pied Oystercatcher	*Haematopus longirostris*
Sooty Oystercatcher	*Haematopus fuliginosus*
Masked Plover	*Vanellus miles*
Black-winged or Pied Stilt	*Himantopus himantopus*
Silver Gull	*Larus novaehollandiae*
Whiskered Tern	*Chlidonias hybrida*
Gull-billed Tern	*Gelochelidon nilotica*
Caspian Tern	*Hydroprogne caspia*
Common Tern	*Sterna hirundo*
Crested Tern	*Sterna bergii*
Laughing Kookaburra	*Dacelo novaeguineae*
Forest Kingfisher	*Halcyon macleayii*
Sacred Kingfisher	*Halcyon sancta*
Mangrove Kingfisher	*Halcyon chloris*
Rainbow-bird	*Merops ornatus*

PASSERINE (PERCHING)

Black-faced Cuckoo-shrike	*Coracina novaehollandiae*
Eastern Whipbird	*Psophodes olivaceus*
Little Wattlebird	*Anthochaera chrysoptera*
Noisy Friarbird	*Philemon corniculatus*
Blue-faced Honeyeater	*Entomyzon cyanotis*
Brown Honeyeater	*Lichmera indistincta*
White-cheeked Honeyeater	*Phylidonyris nigra*
Scarlet Honeyeater	*Myzomela sanguinolenta*

DOLPHIN

Bottlenose	*Tursiops sp.*

WALLABIES

Agile Wallaby	*Macropus agilis*
Red-necked Wallaby	*Macropus rufogriseus*
Whiptail Wallaby	*Macropus parryi*
Swamp Wallaby	*Wallabia bicolor*

PLANT, CREEPER AND SHRUB

Beach Bean	*Canavalia rosea*
Beach Spinifex	*Spinifex sericeus*
Beach Stackhousia	*Stackhousia spathulata*
Blue Waterlily	*Nymphaea gigantea*
Emelies	*Emilia sonchifolia*
Forked Sundew	*Drosera binata*
Red Spatula Sundew	*Drosera spatulata*
Sundew	*Drosera peltata*
Goatsfoot	*Ipomoea pes-caprae*
Pigface	*Carpobrotus glaucescens*
Scented Fan-flower	*Scaveola calendulacea*
Swamp Banksia	*Banksia robur*
Twining Guinea Flower	*Hibbertia scandens*
Wallum Boronia	*Boronia falcifolia*
Wallum Callistemon	*Callistemon pachyphyllus*
Wallum Zieria	*Zieria laxiflora*

TREES

Horsetail or Coastal Oak	*Casuarina equisetifolia sub-species incana*
Forest Oak	*Allocasuarina torulosa*
Swamp Oak	*Casuarina glauca*
Wallum Oak	*Allocasuarina littoralis*
Coastal Banksia	*Banksia integrifolia*
Coastal Paperbark	*Melaleuca quinquenervia*
Pandanus	*Pandanus tectorius*

MANGROVES

Black Mangrove	*Rhizophora stylosa*
Milky Mangrove	*Excoecaria agallocha*
Orange Mangrove	*Bruguiera gymnorhiza*
River Mangrove	*Aegiceras corniculatum*
Spurred Mangrove	*Ceriops tagal var. australis*
White or Grey Mangrove	*Avicennia marina var. australasica*

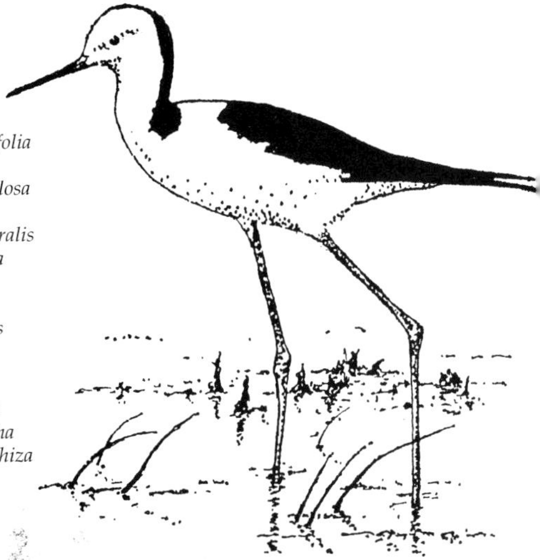

Index of Additional Drawings

Macleay's Swallowtail Butterfly	*Graphium macleayanum macleayanum*	Inside Front Cover
Speckled Anemone	*Oulactis muscosa*	1
Cunjevoi	*Pyura stolonifera*	2
Royal Spoonbill	*Platalea regia*	4
Swamp Lily Orchid	*Phaius tankervilliae*	6
Twining Guinea Flower	*Hibbertia scandens*	7
Black-necked Stork or Jabiru	*Xenorhynchus asiaticus*	8
White-bellied Sea-eagle	*Haliaeetus leucogaster*	9
Caspian Tern	*Hydroprogne caspia*	10
Australian Pelican	*Pelecanus conspicillatus*	10
Common Albatross Butterfly	*Appias paulina ega*	12
Large Banded Awl Butterfly	*Hasora khoda haslia*	13
Blue Nudibranch	*Glaucus spp.*	14
Bêche-de-mer	*Fam. Holothuroidea*	32
Sand Wasp	*Bembix spp.*	36
Mongolian Plover	*Charadrius mongolus*	44
Silver Bream	*Acanthopagrus australis*	49
Chiton or Coat-of-mail Shell	*Liolophura gaimardi*	66
Flag Razor Shell	*Atrina vexillum*	66
Darter or Snake Bird (male)	*Anhinga melanogaster*	81
Head of Krefft's Tortoise	*Emydura kreffti*	92
Striped Marsh Frog	*Limnodynastes peroni*	93
Pied or Black-winged Stilt	*Himantopus himantopus*	94
Mangrove Gerygone	*Gerygone laevigaster*	95
Green Tree Frog	*Litoria caerulea*	96

Photographs by Cecily Fearnley

Habitat drawings and the following additional drawings by David McFarland:

Flying Royal Spoonbills (page 4); Yellow-bellied Sea-snake (page 25); Pandanus on the Cliff (page 46); Tree Goanna (page 61); Forked Sundew (page 85)

Species drawings by Cecily Fearnley

SWAMP BANKSIA PLANT

Grey seed cone and russet aged flower add interest to swamp and windswept heath.

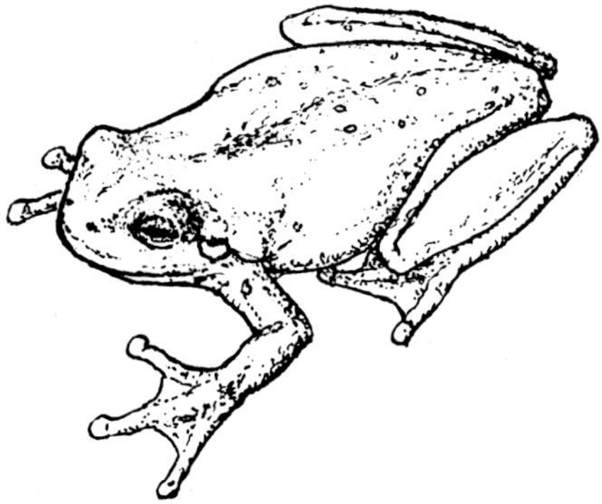

SWAMP BANKSIA FLOWER SPIKE

Buds of the Swamp Banksia are remarkable for their dark green colouring. This changes to a creamy orange when fully open.